T0189402

Lecture Notes in Computer Science 13932

The series Lecture Notes in Computer Science (LNCS), including its subseries Lecture Notes in Artificial Intelligence (LNAI) and Lecture Notes in Bioinformatics (LNBI), has established itself as a medium for the publication of new developments in computer science and information technology research, teaching, and education.

LNCS enjoys close cooperation with the computer science R & D community, the series counts many renowned academics among its volume editors and paper authors, and collaborates with prestigious societies. Its mission is to serve this international community by providing an invaluable service, mainly focused on the publication of conference and workshop proceedings and postproceedings. LNCS commenced publication in 1973.

Hao Chen · Luyang Luo
Editors

Trustworthy Machine Learning for Healthcare

First International Workshop, TML4H 2023
Virtual Event, May 4, 2023
Proceedings

 Springer

Editors
Hao Chen 🆔
Hong Kong University of Science
and Technology
Hong Kong, Hong Kong

Luyang Luo 🆔
Hong Kong University of Science
and Technology
Hong Kong, Hong Kong

ISSN 0302-9743   ISSN 1611-3349 (electronic)
Lecture Notes in Computer Science
ISBN 978-3-031-39538-3   ISBN 978-3-031-39539-0 (eBook)
https://doi.org/10.1007/978-3-031-39539-0

This Springer imprint is published by the registered company Springer Nature Switzerland AG
The registered company address is: Gewerbestrasse 11, 6330 Cham, Switzerland

Preface

We are pleased to present the proceedings of the 1st ICLR Workshop on Trustworthy Machine Learning for Healthcare (TML4H), which was held in a virtual fashion on May 4, 2023.

Machine learning (ML) has achieved or even exceeded human performance in many healthcare tasks, owing to the fast development of ML techniques and the growing scale of medical data. However, ML techniques are still far from being widely applied in practice. Real-world scenarios are far more complex, and ML is often faced with challenges to its trustworthiness such as lack of explainability, generalization, fairness, privacy, etc. Improving the credibility of machine learning is hence of great importance to enhance the trust and confidence of doctors and patients in using the related techniques. The goal of this workshop is to bring together experts from academia, clinical practice, and industry with an insightful vision of promoting trustworthy machine learning in healthcare in terms of scalability, accountability, and explainability.

ICLR 2023 TML4H attracted 30 valid submissions. These submissions were selected through a double-blind peer-review process. We adopted OpenReview for paper submissions.

ICLR 2023 TML4H had a 10-person Program Committee, who were also Area Chairs (AC) responsible for organizing paper reviews. Each area chair was assigned about 3 manuscripts, for each of which they were asked to suggest up to 2–3 potential reviewers. Subsequently, over 58 reviewers were invited. Final reviewer allocations via OpenReview took account of PC suggestions and reviewer bidding, finally allocating about 2 papers per reviewer. Based on the double-blinded reviews, Area Chairs' recommendations, and Program Chairs' overall adjustments, 7 papers (23%) were accepted as long oral presentations, 9 papers (30%) were accepted as short presentations, and 14 papers (47%) were rejected. This process resulted in the acceptance of a total of 16 papers, reaching an overall acceptance rate of 53% for ICLR 2023 TML4H.

We would like to express our gratitude to all of the authors who submitted and presented their outstanding work at TML4H, which made TML4H a resounding success. We are also honoured to have invited Lena Maier-Hein to give the keynote on Trustworthy Machine Learning in Medical Imaging, as well as Minhao Cheng, Huazhu Fu, Shandong Wu, Georgios Kaissis, and Xiaoxiao Li to deliver their insightful talks. We hope these high-quality works and in-depth discussions will pave new ways and inspire new directions in trustworthy ML for healthcare.

June 2023

Hao Chen
Luyang Luo

Organization

Program Committee Chairs

Hao Chen   Hong Kong University of Science and Technology, China

Luyang Luo   Hong Kong University of Science and Technology, China

Program Committee

Yueming Jin	National University of Singapore, Singapore
Jing Qin	Hong Kong Polytechnic University, China
Vince Varut Vardhanabhuti	University of Hong Kong, China
Jiguang Wang	Hong Kong University of Science and Technology, China
Xi Wang	Stanford University, USA
Xin Wang	Chinese University of Hong Kong, China
Daguang Xu	NVIDIA, USA
Yuyin Zhou	University of California, Santa Cruz, USA

Advisory Committee

Marius George Linguraru	Children's National Hospital, USA
Pheng-Ann Heng	Chinese University of Hong Kong, China
Le Lu	Alibaba, China
Danny Z. Chen	University of Notre Dame, USA
Kwang-Ting Cheng	Hong Kong University of Science and Technology, China

Student Organizers

Wenqiang Li   Hong Kong University of Science and Technology, China

Yu Cai   Hong Kong University of Science and Technology, China

Contents

Do Tissue Source Sites Leave Identifiable Signatures in Whole Slide Images Beyond Staining?

Piotr Keller$^{(\boxtimes)}$, Muhammad Dawood, and Fayyaz ul Amir Minhas

University of Warwick, Coventry, UK
Piotr.Keller@warwick.ac.uk

Abstract. *Why can deep learning predictors trained on Whole Slide Images fail to generalize?* It is a common theme in Computational Pathology to see a high performing model developed in a research setting experience a large drop in performance when it is deployed to a new clinical environment. One of the major reasons for this is the batch effect that is introduced during the creation of Whole Slide Images resulting in a domain shift. Computational Pathology pipelines try to reduce this effect via stain normalization techniques. However, in this paper, we provide empirical evidence that stain normalization methods do not result in any significant reduction of the batch effect. This is done via clustering analysis of the dataset as well as training weakly-supervised models to predict source sites. This study aims to open up avenues for further research for effective handling of batch effects for improving trustworthiness and generalization of predictive modelling in the Computational Pathology domain.

Keywords: Stain Normalization · Whole Slide Image · Computational Pathology

1 Introduction

Computational Pathology (CPath) is an emerging field which aims to leverage the ever increasing amount of health data to solve complex and clinically relevant problems through the application of machine learning [2]. Areas of interest include, but are not limited to, predicting diagnostic abnormalities associated with cancer,nuclei instance segmentation and classification [10], cellular composition [6], gene mutations and expression levels [5,7], as well as survival prediction of patients [3,19]. In the recent years, there has been a growing popularity for the use of deep learning methods in CPath which utilise digitised slide tissue images also known as Whole Slide Images (WSIs) [4,17,22]. Such models have been very successful with many studies reporting a multitude of very high performance metrics on a wide range of datasets [21]. Nevertheless, when some of these models are applied in a clinical setting they can fail to generalize [9].

Supported by University of Warwick.

H. Chen and L. Luo (Eds.): TML4H 2023, LNCS 13932, pp. 1–10, 2023.
https://doi.org/10.1007/978-3-031-39539-0_1

In this work we argue that this is partially a consequence of reliance on stain normalization methods in the WSI pre-processing pipeline which are not able to truly remove the variability present in images sourced from different hospitals or laboratories. This variability occurs in the creation of these WSIs. As cells are transparent, it is necessary to stain tissue samples before they are digitised or observed under a microscope to effectively interpret them visually. The staining reagent and process as well as the scanner used can vary across source sites which often results in inconsistent staining characteristics across WSIs. This site-specific signature results in a batch effect which can be exploited by a deep learning model to produce inflated accuracy values but poor generalization in cases where the stain characteristics are different. Stain normalization approaches aim to remove this variability by reducing colour and intensity variations present in these images by normalizing them to a standard or base image. However, as stain normalization usually works in a low-dimensional space, we hypothesise that it fails to remove any higher order site-specific signatures which can still lead to exploitation of the batch effect and generalization failure under domain shift. This means that stain normalized images look normalized to the human eye when in reality hidden factors such as those that result from different laboratory protocols are still present. These factors can skew the learning process acting as confounding variables which could lead to an overestimation of a model's true performance on a given task and subsequently be the cause of the poor generalisation in clinical deployment.

The technical contribution of this paper is the demonstration of empirical evidence of the presence of these hidden factors in a dataset regardless of what stain normalization technique is applied to it. This is achieved through a carefully designed experiment using different stain normalization schemes as well as two fundamentally different types of predictors. This aspect of the design of CPath pipelines is often ignored and, to the best of our knowledge, has not been extensively explored. We have only found one instance where the phenomenon of source site signatures was carefully explored with a carefully controlled experiment in the current literature [12]. Their work focused on using patch level classification to predict source sites, concluding that source site signatures persist even after stain normalisation. Our work extended this and to the best of our knowledge is the first instance to study this phenomenon on a whole slide level. The findings in this paper have significant bearing on improving trustworthiness and generalization of machine learning applications in the rapidly emerging area of CPath.

2   Materials and Methods

2.1   Experiment Design Strategy

To illustrate that stain normalization methods are unable to effectively remove center specific batch effects, we designed a simple experiment in which we predict the laboratory of origin (centre) of a WSI both before and after stain normalization. We first predict the centre of origin of a WSI by modelling this task

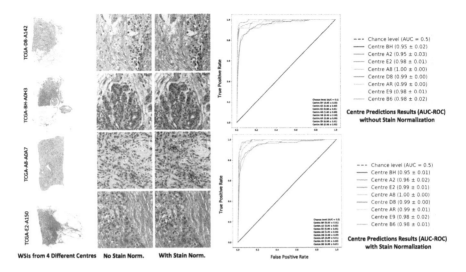

Fig. 1. The left of the figure shows a visualisation of four WSIs originating from different source sites to highlight the staining variations found across source sites. For each WSI we show an example non-stain normalized patch and also a corresponding Macenko stain-normalized version to visualise how stain normalization appears to remove staining variation. The right of the figure, ROC cures are displayed for predicting tissue source site of WSIs using both non-stain normalized and stain-normalized images. The ROC curves are generated using an SVM that utilizes a pre-computed Maximum Mean Discrepancy kernel computed with ShuffleNet features [14].

as a weakly-supervised binary classification problem with the target label being the centre of origin. We then develop a separate weakly-supervised predictor to predict the centre of origin with stain normalized WSIs as input. The fundamental principle behind this experiment is that if stain normalization is an effective strategy to remove any identifiable signatures of the centre of origin or the underlying batch effect, we should get substantial decrease in accuracy of predicting the center after stain normalization. For this purpose, we use multicentric breast cancer WSIs from the Cancer Genome Atlas (TCGA) [1]. To show that the results of our analysis are not specific to a certain type of stain normalization, we utilize two different yet commonly used stain normalization schemes (Reinhard [20] and Macenko [18]). In order to marginalize the effect of the choice of the weakly supervised method being used for the prediction of the centre of origin, we use two fundamentally different types of predictors (CLAM [16] and MMD-Kernels [13]). Below each component of the experiment design is explained in further detail.

2.2 Dataset

1,113 publicly available WSIs of Formalin-Fixed paraffin-Embedded (FFPE) Hematoxylin and Eosin (H&E) stained tissue sections of 1084 breast carcinoma patients were collected from The Cancer Genome Atlas (TCGA-BRCA) [1,11]. For some patients multiple WSIs were available and thus only the ones with best visual quality were used. Additionally, WSIs with missing baseline resolution information were ignored. After filtering 1,051 WSIs remain which are used for analysis. These WSIs were belonging to 49 sources sites.

2.3   Pre-processing of WSIs

Quality of WSIs can be negatively affected by artefacts (tissue folds, pen-marking, etc.) initiating from histology laboratories. To ensure that any models do not exploit these tissue artefacts the tissue regions of WSIs are segmented using a tissue segmentation method. The tissue segmentation means that only information tissue regions remain and artefacts are removed. Since, an entire WSI at full resolution can be very large ($100,000 \times 100,000$ pixels) and cannot be fitted into a GPU memory each WSI is tiled into patches of size 512×512 at a spatial resolution of 0.50 microns-per-pixel (MPP). Tiles that capture less than 40% of informative tissue area (mean pixel intensity greater than 200) are filtered out.

2.4   Tissue Staining and Stain Normalization Methods

Histology images are acquired by staining a tissue specimen with a dye that shows variable affinities to different tissue components. In case of routine Hematoxylin and Eosin (H&E) staining, nuclei are stained with Hematoxylin and are highlighted in bluish color, while cytoplasm and extracellular matrix are stained with Eosin and can be seen in pinkish color [8]. However, due to variations in staining protocols, characteristics of the dye, duration for which the dyes are applied, tissue type and thickness, scanner characteristics and a number of other factors can impact the stain characteristics of the tissue resulting in center-specific confounding factors which are not at all related to any underlying pathology. These constitute a batch effect that can leave a centre-specific signature in the tissue image and affect the generalization performance of any machine learning method.

One way of addressing such variations is stain normalization using methods such as the ones proposed by Reinhard [20] and Macenko [18]. These stain normalization methods map the color style of source image to target images [15] while preserving cellular and morphometric information contained in the images.

2.5   Source Site Prediction

We hypothesise that majority of stain normalization methods try to make the images look similar but even after stain-normalization the histology laboratory from which tissue specimen is originating can still be predicted. More specifically, we argue that the use of stain normalization methods are not likely to make CPath algorithms generalize in case of domain-shift as these methods can not completely eliminate stain-specific information of the source site. To illustrate this, we used stain-normalized and non stain-normalized images and tried to predict the tissue source site as target variable. The hypothesis is that, if stain-normalization is removing the tissue site specific information then the tissue source site should be significantly less predictable from the stain-normalized images compared to non-stain-normalized images.

We demonstrated the predictability of tissue source site from stain-normalized and non-stain normalized images using a multiple instance learning method and also a kernel based method. As a multiple instance learning method, we used Clustering-constrained Attention Multiple Instance Learning (CLAM) which is a weakly-supervised method that has shown promising performance in several computational pathology tasks [16]. CLAM considers each WSI as a bag of patches and then used attention-based pooling function for obtaining slide-level representation from patch-level representation. As a second predictive model, we used a recently published support vector machine (SVM) based classification method that constructs a whole slide level kernel matrix using Maximum Mean Discrepancy (MMD) over ShuffleNet derived feature representations of patches in WSIs [13]. In recent work, this method has been shown to have strong predictive power for TP53 mutation prediction and survival analysis from WSIs. Note that the two methods have fundamentally different principles of operation so that any subsequent findings can be understood in a broad context independent of the specific nature of the predictive model being used. As it is not the goal of this work to present these specific predictors, the interested reader is referred to their original publications for further details.

We evaluated the performance of both these methods in predicting tissue source site using both stain-normalized and non-stain normalized data. The experiments were performed using stratified five fold cross validation. For each source site we train a separate model using one-vs-rest approach, in which all tissue images of patients originating from a given source site L are labelled as 1, while the rest are labelled as 0. We then train the predictive model for predicting the source site of each WSI. In order to make meaningful comparisons, we restricted our analysis to prediction of 8 sources sites each of which has 50 or more images each.

The hyper-parameters were selected by utilising a validation set (30% of each train split fold). Average Area under the Receiver Operating Characteristic curve (AUCROC) across the 5 folds along with its standard deviation was used as the performance metric.

2.6   Similarity Kernel and Clustering Analysis

In order to further understand the implication of stain normalization at a dataset level, we performed hierarchical clustering over the WSI MMD kernel matrix for the whole dataset. The matrix shows the degree of pairwise similarity between WSIs. We show the kernel matrices both before and after stain normalization together with clustering. If the stain normalization had been effective at removing any information about the center, we would expect that any clustering done after stain normalization will not be possible to group WSIs from the same center into the same cluster. This serves as an additional un-supervised analysis of whether clustering is able to remove center-specific information or not.

3   Results and Discussion

3.1   Effect of Stain Normalization

Figure 1 shows the visual results of applying stain normalization to a patch belonging to 4 example WSIs each originating from a different centre. From the figure it can be clearly seen that patches belonging to different centers look the same after normalization hence to the human eye it would seem that we have removed the batch effect. However looking at the ROC curves we can see that both before and after stain normalization MMD kernels can near perfectly distinguish the WSI origin. This supports our hypothesis that stain-normalization methods are not removing the source site information. Even if after stain normalization the WSIs look the same the underlying footprint is still there. If stain normalization methods have truly removed source site information, then we will be seeing AUCROC of 0.5 (random) but this is not the case. From this analysis we can say that, the analyzed stain-normalization methods are less likely to make models robust against domain-shift.

3.2   Predictive Power over Original Data

Tables 1–2 show the results of prediction of the source center from original WSIs, i.e., without any stain normalization using two different predictive pipelines (CLAM in Table-1 and MMD Kernel in Table-2). These results show that it is possible to predict the source of a given WSI with very high predictive power as measured using AUCROC for both methods. This shows that, as expected, there is a significant signature in a WSI specific to the laboratory of origin.

3.3   Predictive Power over Stain Normalized Data

Tables 1–2 show the results of prediction of the source center from stain normalized WSIs, i.e., with stain normalization using two different predictive pipelines (CLAM in Table-1 and MMD Kernel in Table-2) and two different stain normalization methods (Reinhard and Macenko stain normalization). These results show that it is possible to predict the source of a given WSI with very high predictive power as measured using AUCROC using both predictive pipelines even after stain normalization. There is effectively very little change in predictive power as a consequence of stain normalization. This shows that stain normalization alone is not able to remove the site-specific information contained in a WSI and the batch effect still exists even after stain adjustment.

3.4   Clustering Analysis

The hierarchically-clustered heatmaps along with their respective dendrograms for the kernels are shown in Fig. 2. From this figure one can see that both non-normalized and stain normalized WSIs, have a large proportion of brightly coloured regions in their heatmaps indicating that there are many slides that share similar characteristics. The dataset has been split into 4 main clusters as can be seen on the dendrograms where slides within the same cluster seem to regularly originate from the same laboratory, for example the orange cluster contains many slides from laboratory E2 (Roswell Park). This indicates to us that some hidden site identification markers are likely to still be present even after normalization.

3.5   Code and Data Availability

The code and data used in this paper are publicly available on GitHub.

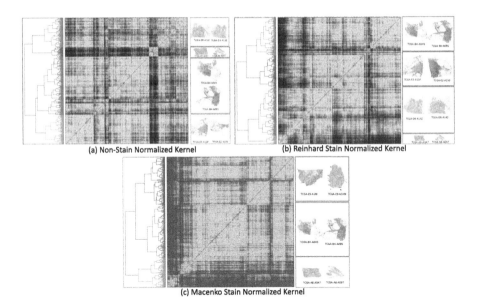

(a) Non-Stain Normalized Kernel   (b) Reinhard Stain Normalized Kernel

(c) Macenko Stain Normalized Kernel

Fig. 2. The visualisation of the kernels obtained by computing the MMD kernel on whole TCGA Breast cancer cohort. a) Show the pairwise similarity of WSI using non-stain normalized data, while b) and c) show the kernel matrices for Reinhard and Macenko stain normalized WSIs respectively. Each kernel also has an associated dendrogram as well as a visualisation of some of the patches from each of its clusters.

Table 1. Performance of CLAM (with standard deviation, σ) trained for source site prediction for various stain normalization protocols. Here $+, -$ indicate WSIs that originated from the chosen site and from one of the remaining source sites respectively.

Source Site	$(+, -)$	Unnormalized AUCROC $\pm \sigma$	Reinhard AUCROC $\pm \sigma$	Macenko AUCROC $\pm \sigma$
University of Pittsburgh (BH)	(142, 903)	0.84 ± 0.04	0.82 ± 0.06	0.86 ± 0.03
Walter Reed (A2)	(100, 945)	0.82 ± 0.10	0.73 ± 0.07	0.87 ± 0.07
Roswell Park (E2)	(90, 955)	0.96 ± 0.01	0.92 ± 0.02	0.96 ± 0.01
Indivumed (A8)	(74, 971)	1.00 ± 0.00	0.99 ± 0.00	1.00 ± 0.00
Greater Poland Cancer Center (D8)	(78, 967)	0.97 ± 0.03	0.94 ± 0.04	0.97 ± 0.02
Mayo (AR)	(69, 976)	0.98 ± 0.02	0.95 ± 0.04	0.97 ± 0.03
Asterand (E9)	(62, 983)	0.98 ± 0.02	0.98 ± 0.01	0.96 ± 0.04
Duke (B6)	(50, 995)	0.97 ± 0.03	0.92 ± 0.04	0.94 ± 0.06
Average AUCROC		0.94 ± 0.08	0.94 ± 0.06	0.91 ± 0.09

Table 2. Performance of MMD Kernels (with standard deviation, σ) trained for source site prediction for various stain normalization protocols. Here $+, -$ indicate WSIs that originated from the chosen site and from one of the remaining source sites respectively.

Source Site	$(+, -)$	Unnormalized AUCROC $\pm \sigma$	Reinhard AUCROC $\pm \sigma$	Macenko AUCROC $\pm \sigma$
University of Pittsburgh (BH)	(142, 903)	0.95 ± 0.02	0.93 ± 0.02	0.95 ± 0.01
Walter Reed (A2)	(100, 945)	0.95± 0.03	0.88 ± 0.04	0.96 ± 0.02
Roswell Park (E2)	(90, 955)	0.98 ± 0.01	0.98 ± 0.02	0.99 ± 0.01
Indivumed (A8)	(74, 971)	1.0 ± 0.00	1.0 ± 0.00	1.0 ± 0.00
Greater Poland Cancer Center (D8)	(78, 967)	0.99 ± 0.00	0.99 ± 0.01	0.99 ± 0.00
Mayo (AR)	(69, 976)	0.99 ± 0.00	0.98 ± 0.01	0.99 ± 0.01
Asterand (E9)	(62, 983)	0.98 ± 0.01	0.98 ± 0.01	0.98 ± 0.02
Duke (B6)	(50, 995)	0.98 ± 0.02	0.97 ± 0.01	0.98 ± 0.01
Average AUCROC		0.98 ± 0.02	0.96 ± 0.04	0.98 ± 0.02

4   Conclusions and Future Work

We conclude that tissue source sites leave identifiable markers that can be picked by machine learning models. We show that this may be one of the reasons why many models often result in poor generalization when used outside a research setting thus we urge computational pathologists to keep this in mind when designing models and datasets. In the future we would like to verify our results on a larger database as well as explore what exactly are the most prominent factors that make a source site so easily distinguishable and how we can develop strategies to counter such confounding factors. On top of this we would like to see if there is a performance change when a model takes into account these source site signatures during training.

Acknowledgements. FM acknowledges funding from EPSRC EP/W02909X/1 and PathLAKE consortium. MD and FM report research funding from GlaxoSmithKline outside the submitted work.

References

1. Liu, J., et al.: An integrated TCGA pan-cancer clinical data resource to drive high-quality survival outcome analytics. Cell **173**(2), 400–416.e11 (2018). https://doi.org/10.1016/j.cell.2018.02.052
2. Abels, E., et al.: Computational pathology definitions, best practices, and recommendations for regulatory guidance: a white paper from the digital pathology association. J. Pathol. **249**(3), 286–294 (2019)
3. Chen, R.J., et al.: Scaling vision transformers to gigapixel images via hierarchical self-supervised learning. In: Proceedings of the IEEE/CVF Conference on Computer Vision and Pattern Recognition, pp. 16144–16155 (2022)
4. Chen, R.J., et al.: Pan-cancer integrative histology-genomic analysis via multimodal deep learning. Cancer Cell **40**(8), 865–878 (2022)
5. Coudray, N., et al.: Classification and mutation prediction from non-small cell lung cancer histopathology images using deep learning. Nat. Med. **24**(10), 1559–1567 (2018)
6. Dawood, M., Branson, K., Rajpoot, N.M., Minhas, F.: Albrt: cellular composition prediction in routine histology images. In: Proceedings of the IEEE/CVF International Conference on Computer Vision, pp. 664–673 (2021)
7. Dawood, M., Branson, K., Rajpoot, N.M., Minhas, F.U.A.A.: All you need is color: image based spatial gene expression prediction using neural stain learning. In: Kamp, M., et al. (eds.) Machine Learning and Principles and Practice of Knowledge Discovery in Databases. ECML PKDD 2021. CCIS, vol. 1525, pp. 437–450. Springer, Cham (2021). https://doi.org/10.1007/978-3-030-93733-1_32
8. Fischer, A.H., Jacobson, K.A., Rose, J., Zeller, R.: Hematoxylin and eosin staining of tissue and cell sections. Cold Spring Harb. Protoc. **2008**(5), pdb-prot4986 (2008)
9. Foote, A., Asif, A., Rajpoot, N., Minhas, F.: REET: robustness evaluation and enhancement toolbox for computational pathology. Bioinformatics **38**(12), 3312–3314 (2022)
10. Graham, S., et al.: Lizard: a large-scale dataset for colonic nuclear instance segmentation and classification. In: Proceedings of the IEEE/CVF International Conference on Computer Vision, pp. 684–693 (2021)
11. Hoadley, K.A., et al.: Cell-of-origin patterns dominate the molecular classification of 10,000 tumors from 33 types of cancer. Cell **173**(2), 291–304 (2018)
12. Howard, F.M., et al.: The impact of site-specific digital histology signatures on deep learning model accuracy and bias. Nat. Commun. **12**(1), 4423 (2021)
13. Keller, P., Dawood, M., Minhas, F.U.A.A.: Maximum mean discrepancy kernels for predictive and prognostic modeling of whole slide images. arXiv preprint arXiv:1111.6285 (2023)
14. Keller, P., Dawood, M., et al.: Maximum mean discrepancy kernels for predictive and prognostic modeling of whole slide images. arXiv preprint arXiv:2301.09624 (2023)

15. Khan, A.M., Rajpoot, N., Treanor, D., Magee, D.: A nonlinear mapping approach to stain normalization in digital histopathology images using image-specific color deconvolution. IEEE Trans. Biomed. Eng. **61**(6), 1729–1738 (2014)
16. Lu, M.Y., Williamson, D.F., Chen, T.Y., Chen, R.J., Barbieri, M., Mahmood, F.: Data-efficient and weakly supervised computational pathology on whole-slide images. Nat. Biomed. Eng. **5**(6), 555–570 (2021)
17. Lu, W., Toss, M., Dawood, M., Rakha, E., Rajpoot, N., Minhas, F.: Slidegraph+: whole slide image level graphs to predict her2 status in breast cancer. Med. Image Anal. **80**, 102486 (2022)
18. Macenko, M., et al.: A method for normalizing histology slides for quantitative analysis. In: 2009 IEEE International Symposium on Biomedical Imaging: From Nano to Macro, pp. 1107–1110. IEEE (2009)
19. Mackenzie, C.C., Dawood, M., Graham, S., Eastwood, M., ul Amir Afsar Minhas, F.: Neural graph modelling of whole slide images for survival ranking. In: Rieck, B., Pascanu, R. (eds.) Proceedings of the First Learning on Graphs Conference. Proceedings of Machine Learning Research, vol. 198, pp. 48:1–48:10. PMLR, 09–12 December 2022
20. Reinhard, E., Adhikhmin, M., Gooch, B., Shirley, P.: Color transfer between images. IEEE Comput. Graph. Appl. **21**(5), 34–41 (2001)
21. Sokolova, M., Japkowicz, N., Szpakowicz, S.: Beyond accuracy, F-Score and ROC: a family of discriminant measures for performance evaluation. In: Sattar, A., Kang, B. (eds.) AI 2006. LNCS (LNAI), vol. 4304, pp. 1015–1021. Springer, Heidelberg (2006). https://doi.org/10.1007/11941439_114
22. Yao, J., Zhu, X., Jonnagaddala, J., Hawkins, N., Huang, J.: Whole slide images based cancer survival prediction using attention guided deep multiple instance learning networks. Med. Image Anal. **65**, 101789 (2020)

Explaining Multiclass Classifiers with Categorical Values: A Case Study in Radiography

Luca Franceschi[1,2](✉), Cemre Zor[1,2], Muhammad Bilal Zafar[1,2],
Gianluca Detommaso[1,2], Cedric Archambeau[1,2], Tamas Madl[1,2], Michele Donini[1,2],
and Matthias Seeger[1,2]

[1] Amazon Web Services, Berlin, Germany
{franuluc,cemrezor}@amazon.co.uk
[2] Amazon Web Services, London, UK

Abstract. Explainability of machine learning methods is of fundamental importance in healthcare to calibrate trust. A large branch of explainable machine learning uses tools linked to the Shapley value, which have nonetheless been found difficult to interpret and potentially misleading. Taking multiclass classification as a reference task, we argue that a critical issue in these methods is that they disregard the structure of the model outputs. We develop the Categorical Shapley value as a theoretically-grounded method to explain the output of multiclass classifiers, in terms of transition (or flipping) probabilities across classes. We demonstrate on a case study composed of three example scenarios for pneumonia detection and subtyping using X-ray images.

1 Introduction

Machine learning (ML) has emerged as a powerful tool in healthcare with the potential to revolutionize the way we diagnose, treat and prevent diseases. ML algorithms have a wide range of applications including early detection of diseases, risk prediction in patients developing certain conditions, optimisation of treatment plans, improved prognosis, assistance in clinical decision-making, gene expression analysis, genomic classification, improved personalize patient care and more. However, the adoption of ML in clinical practice has often been hampered by the opaqueness of the ML models. This opaqueness may trigger skepticism in clinicians and other end-users such as patients or care-givers to trust model recommendations without understanding the reasoning behind their predictions, which delays and/or decreases the adoption of state-of-the art technologies and hinders further advances.

Various methods have been proposed in the literature to enhance the explainability of ML models (XAI). Among these, (local) feature attribution methods such as SHAP (Lundberg and Lee, 2017) or variants (e.g. Frye et al., 2020, Chen et al., 2018, Heskes et al., 2020) have gained considerable traction. In fact, Shapley value based explanations are the most popular explainability methods according to a recent study by Bhatt et al. (2020) These methods, supported by a number of axioms (properties) such as nullity, linearity and efficiency, provide insight into the contribution of each feature towards

© The Author(s), under exclusive license to Springer Nature Switzerland AG 2023
H. Chen and L. Luo (Eds.): TML4H 2023, LNCS 13932, pp. 11–24, 2023.
https://doi.org/10.1007/978-3-031-39539-0_2

the model decisions. There is, however, a growing scrutiny into the utility of these techniques, which are judged to be unintuitive and potentially misleading (Kumar et al., 2020, Mittelstadt et al., 2019), and do not support contrastive statements (Miller, 2019). While part of these issues may be rooted in misinterpretations of the technical tools involved[1], in this paper we argue that a critical flaw in current approaches is their failure to capture relevant structure of the object one wishes to explain (the explicandum). In contrast, we take the position that attributive explanations should comply with the nature of the explicandum: in particular, if the model output is a random variable (RV), we should represent marginal contributions as RVs as well. Our contribution, which we dub *the Categorical Shapley value*, can fully support statements such as "the probability that the feature x_i causes x to be classified as viral pneumonia rather than bacterial pneumonia is y", which we develop, experiment and discuss in this paper within the context of X-ray imaging.

1.1 The Shapley Value and Its Application to Explain Multiclass Classifiers

For concreteness, we focus here on the multiclass classification (d classes) as one of the most common tasks in ML. Let $f : X \subseteq \mathbb{R}^n \mapsto \mathcal{Y}$ be a (trained) multiclass classifier and $x \in X$ an input point. One common strategy to explain the behaviour of the model at x is to attribute an importance score to each input feature through the computation of the Shapley value (SV) (Shapley, 1953a). In order to do so, one must first construct a cooperative game v where players correspond to features, and coalitions correspond to features being used: that is $v(S) = f(x_{|S})$, where $S \in 2^n$.[2] Then, for each $i \in [n]$, the Shapley value is given by

$$\psi_i(v) = \sum_{S \in 2^{[n]\backslash i}} p(S)[v(S \cup i) - v(S)] = \mathbb{E}_{S \sim p(S)}[v(S \cup i) - v(S)]; \qquad (1)$$

where $p(S) = \frac{1}{n}\binom{n-1}{|S|}^{-1}$ if $i \notin S$ and 0 otherwise. The quantity $v(S \cup i) - v(S)$ is called the marginal contribution of i to the coalition S. See Roth (1988) for an in-depth discussion of the SV and related topics.

Historically, the SV has been developed as an answer to the question: How can we assign a worth (or value) to each player i? The SV does so by distributing "fairly" the *grand payoff* $v([n])$ among players, so that (1) if a player never contributes to the payoff, their worth is null, (2) if any two players have indistinguishable marginal contributions, they have the same worth, and (3) if v is a linear combination of two games, say u and w, then the worth of i for v is the corresponding linear combination of their worth for u and w. The game v could typically represent an economic or political process (e.g. a vote) and, critically, would be modelled as a real-valued set function; i.e. $v : 2^d \mapsto \mathbb{R}$, so that $\psi_i(v) \in \mathbb{R}$.

---

[1] For instance, the Shapley value is a descriptive rather than prescriptive tool. This means that, in general, one should not expect that changing the feature with the highest Shapley value should lead to the largest change in the outcome.

[2] In practice, out-of-coalition features must often be given a value; this could be an arbitrary baseline, a global or a conditional average Sundararajan and Najmi (2020), Aas et al. (2021).

2   Categorical Games and Values

In our case, the grand payoff is the output $f(x)$ that determines the class the model assigns to x. Whilst in practice f could be implemented in various ways, several modern ML models (e.g. neural networks) output *distributions* over the classes – e.g. through a softmax layer. Equivalently, one may think of $f(x)$ as an E-valued (categorical) random variable. Using the one-hot-encoding convention, we identify $E = \{e_s\}_{s=1}^d$ as the one-hot vectors of the canonical base of $\mathbb{R}^d$. Now, however, it becomes unclear which real number should be assigned to a difference of random variables. Moreover, averaging over coalitions S, as done in Eq. (1), may induce a semantic gap in this context. To recover the standard pipeline to compute the SV, one may settle for explaining the logits or the class probabilities as if they were independent scalars. However, this may lead to paradoxical explanations that attribute high importance to a certain feature (say x_1) for *all* classes, failing to capture the fact that an increase in the likelihood of a given class must necessarily result in an aggregated decrease of the likelihood of the others. Here we show how to avoid such step which causes loss of structure and rather explain $f(x)$ directly as a random variable.

For a player i and a coalition S not containing i, we need to relate $v(S)$ with $v(S \cup i)$ in order to quantify the marginal contribution of i to S. This relationship is not just in terms of the marginal distributions of these two variables, but also of their dependence. In this paper, we assume a simple dependency structure between all variables $v(S)$, in that $v(S) = \tilde{v}(S, \varepsilon)$ for $\varepsilon \sim p(\varepsilon)$ where $\tilde{v}$ is a deterministic mapping to E, and ε is a random variable distributed according to some $p(\varepsilon)$. Let $v(S)$ be a d-way categorical distribution with natural parameters $\theta_{S,j}$, in that

$$\mathbb{P}(v(S) = j) = \frac{e^{\theta_{S,j}}}{\sum_k e^{\theta_{S,k}}} = \text{Softmax}(\theta_S).$$

We call such v a *Categorical game*. We can implement the aforementioned dependency assumption by the Gumbel-argmax reparameterization (Papandreou and Yuille, 2011): $\tilde{v}(S, \varepsilon) = \arg\max_k\{\theta_{S,k} + \varepsilon_k\}$, where $\varepsilon_1, \ldots, \varepsilon_d$ are independent standard Gumbel variables.

Given this construction, we redefine the *marginal contribution* of i to S as the random variable $\tilde{v}(S \cup i, \varepsilon) - \tilde{v}(S, \varepsilon)$ for $\varepsilon \sim p(\varepsilon)$. This RV takes values in the set $E - E = \{e - e' \mid e, e' \in E\}$; we shall call its distribution

$$q_{i,S}(z) = \mathbb{P}(v(S \cup i) - v(S) = z \mid S), \quad z \in E - E.$$

Note that $q_{i,S}(x)$ is a conditional distribution, given $S \in 2^{[n]\setminus i}$ and $E - E$ is a set containing $0 \in \mathbb{R}^d$ and all vectors that have exactly two non-zero entries, one with value $+1$ and the other -1.

We can view this construction as a generalized difference operation $v(S \cup i) \ominus v(S)$ between random variables rather then deterministic values, where the $\ominus$ incorporates the above dependency assumption. We define our *Categorical Shapely value* as the random variable $\xi(v) = \{\xi_i\}_{i \in [n]}$, where

$$\xi_i(v) = v(S_i \cup i) \ominus v(S_i) = \tilde{v}(S \cup i, \varepsilon) - \tilde{v}(S, \varepsilon) \qquad \text{for } \varepsilon \sim p(\varepsilon) \text{ and } S \sim p(S). \quad (2)$$

Note these RVs have multiple sources of randomness, which are independent from each other. We can marginalise over $p(S)$ to obtain the distribution $q_i(x)$ of $\xi_i(v)$: for every $z \in E - E$:

$$q_i(z) = \mathbb{P}(\xi_i(v) = z) = \mathbb{E}_{S_i \sim p^i}[q_{S_i,i}(z)] = \sum_{S_i \in 2^{[n]\backslash i}} p(S_i) q_{S_i,i}(z). \tag{3}$$

One major advantage of this novel construction is that now the distribution of the Categorical SV is straightforward to interpret. Indeed, the probability masses at each point $z = e_r - e_s \in E - E$ are interpretable as the probability (averaged over coalitions) that player i causes the payoff of v (and hence the prediction of f to flip from class s to class r. We refer to $q_i(e_r - e_s)$ as the *transition probability* induced by feature i.

Interestingly, we can derive a closed form analytical expression for the $q_{i,S}$ and, hence, for the q_i. We do this in Appendix A. The following proposition relates the Categorical Shapley value with the standard SV and gives a number of properties that can be derived for the categorical SV.

Proposition 1. *Let ξ be the Categorical Shapley value defined in Eq. (2). Then:*

1. $\mathbb{E}[\xi_i(v)] = \psi_i(\mathbb{E}[v]) \in [-1, 1]^d$, *where $\mathbb{E}[v]$ is the n-players game defined as $\mathbb{E}[v](S) = \mathbb{E}[v(S)] = \mathrm{Softmax}(\theta_S)$;*
2. *If i is a null player, i.e. $v(S \cup i) = v(S)$ for all $S \neq \emptyset$, then $\xi_i(v) = \delta_0$, where δ_0 is the Dirac delta centered in $0 \in \mathbb{R}^d$;*
3. *If $v = v'$ with probability $\pi \in [0, 1]$ and $v = v''$ with probability $1 - \pi$ (independent from S), then $q_i(z) = \mathbb{P}(\xi_i(v) = z) = \pi\mathbb{P}(\xi_i(v') = z) + (1 - \pi)\mathbb{P}(\xi_i(v'') = z) = \pi q_i'(z) + (1 - \pi)q_i''(z)$.*
4. $v([n]) \ominus v(\emptyset) = \sum_{i \in [n]} \mathbb{E}_{S \sim p(S)}[\xi_i(v)]$, *where the sum on the right hand side is the sum of (dependent) $E - E$-valued random variables.*

Property 1 essentially shows that the Categorical SV is strictly more expressive than the traditional Shapley values, whilst putting the traditional SVs for multiclass classification under a new light. Properties 2, 3, and 4 may be seen as the "adaptations" to the Categorical SV of the null player, linearity and efficiency axioms, respectively. In particular, the standard linearity axiom would be of little consequence in this context as taking a linear combination of categorical RVs does not lead to another categorical RV. Instead, Property 3 addresses the common situation where the classifier one wishes to explain is a (probabilistic) ensemble, relating the distributions of the respective Categorical SVs. See Appendix C for a brief discussion of the related work in the cooperative game theory literature.

3   Detecting Pneuomonia in Chest X-Rays: A Case Study

This section employs the Categorical SV (CSV) to analyse a commonly used deep learning architecture, ResNet-18 (He et al., 2015) for pneumonia detection and subtyping using X-ray images, which is casted as a multiclass classification problem based categorising subjects into three classes: healthy controls (HC - class 0), bacterial pneumonia cases (BP - class 1) and viral pneumonia cases (VP - class 2). The model has been

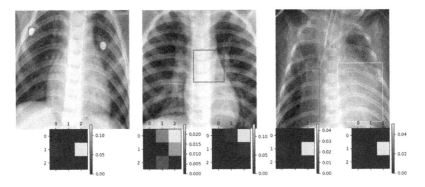

Fig. 1. Three example subject X-ray images and Categorical Shapley values relative to the depicted patches, plotted as matrices. (Left) Ground-truth: VP Prediction: BP. Patch representing two artifacts which should not impact the model decision (Center) Ground-truth: VP Prediction: VP. Two patches, the red one on the left highlighting a section where pneumonia is visible and blue depicting a patch in the middle mediastinum. (Right) Ground-truth: VP Prediction: BP. The red patch is related to a pneumonia area, the yellow one highlights the heart region the patient. (Color figure online)

trained on chest X-ray images collected from pediatric patients, aged one to five, as part of their routine clinical care in Guangzhou Women and Children's Medical Center (Kermany et al., 2018). The aim is to show the importance of using structured explanations even when the model is fine-tuned to the problem of interest, in this case with a mean balanced accuracy score of 84.7%. We select three example scenarios (as depicted in Fig. 1) to analyse different use-cases where CSV empowers the decision process.

Case One: Artifacts. Figure 1 (Left) shows an example scenario of an image with artifacts as depicted in red bounding boxes. The probability distribution output by the model for the ground-truth class BP and the predicted class VP are given as 47.8% and 48.0% respectively. CSV measures the transition probability from VP to BP, generated by the artifact regions, as 12.7%, which implies the presence of these artifacts as a root cause behind the confusion between BP and VP.

Case Two: Correct Classification. Figure 1 (Center) shows a correctly classified VP. However, even though the main affected area in this patient is depicted by red by independent experts, the contribution of this area to the decision has been found negligible (around 1%, see the left matrix under the Center image), making the model's recommendation untrustworthy. Furthermore, the transition probability from VP to HC calculated for the middle mediastinum region (depicted in blue), which is not expected to be a region of interest for pneumonia, can found as high as 13.3%, flagging this region as incorrectly important for the decision process of the model.

Case Three: Incorrect Classification. When the incorrectly classified case shown in Fig. 1 (Right) is analysed, the transition probability for the area in red, which is labelled as a main affected area of VP by independent experts, from the prediction class BP to the ground-truth class VP is calculated as zero. On the other hand, the heart region identified by yellow is shown to exhibit over 5% transition probability to the VP and

BP classes, although this value would be expected to be close to zero. Both of these findings help highlight inconsistencies in the behaviour of the model.

4   Discussion and Conclusion

By analysing three example scenarios in Sect. 3, we have underlined the importance of using model explainability even for fine-tuned, seemingly highly performing models, especially for use in critically important application areas such as healthcare. Employing categorical games and values empowers a structural understanding of the multiclass classification problem by providing information about transition probabilities across classes informing about "decision flips", in addition to the feature contribution information obtained from classical methods. Such knowledge would highly strengthen the model design process; e.g. by promoting the use of comprehensive pre-processing steps, ensemble classification designs or intelligent model tuning.

While we implement a case study on pneumonia classification using X-ray images as a proof-of-concept, the method proposed is extendable to all modalities including genomics, free-text or tabular data. For out of coalition portions of the image, we employed a simple background constant value. We plan to consider more sophisticated formulations in the future. Another invaluable path for future work includes developing better visualization and summarization methods as well as interactive interfaces to support clinicians and other end-users.

A   Analytic Expression of the PDF of Categorical Differences

Consider $E = \{e_1, \ldots, e_d\}$ with $d \geq 3$. Suppose that $v(S)$ has a d-way categorical distribution with natural parameters $\theta_{S,j}$, in that

$$\mathbb{P}(v(S) = j) = \frac{e^{\theta_{S,j}}}{\sum_k e^{\theta_{S,k}}}.$$

Categorical games emerge, e.g., when explaining the output of multiclass classifiers or attention masks of transformer models (Kim et al., 2017, Vaswani et al., 2017).

A latent variable representation is given by the Gumbel-argmax reparameterization (Papandreou and Yuille, 2011):

$$\tilde{v}(S, \varepsilon) = \arg \max_k \{\theta_{S,k} + \varepsilon_k\},$$

where $\varepsilon_1, \ldots, \varepsilon_d$ are independent standard Gumbel variables with probability distribution function $p(\varepsilon_j)$ and cumulative distribution function $F(\varepsilon_j)$ given by

$$F(\varepsilon_j) = \exp\left(-e^{-\varepsilon_j}\right), \quad p(\varepsilon_j) = \exp\left(-\varepsilon_j - e^{-\varepsilon_j}\right).$$

At this point, assume that $e_j = [\mathbf{1}_{k=j}]_k \in \{0, 1\}^d$ are the standard basis vectors of $\mathbb{R}^d$. Then, $E - E = \{e_r - e_s \mid 1 \leq r, s \leq d\}$ has size $d^2 - d + 1$, and the distribution of $v(S \cup i) \ominus v(S)$ is given by the off-diagonal entries of the joint distribution $Q_{i,S}(r, s) = \mathbb{P}(v(S \cup i) = r, v(S) = s)$.

We can work out $Q_{i,S}(r, s)$ explicitly. Denote

$$\alpha_j = \theta_{S \cup i, j}, \quad \beta_j = \theta_{S, j}, \quad \rho_j = \alpha_j - \beta_j.$$

Without loss of generality, we assume the categories to be ordered so that $\rho_1 \geq \rho_2 \geq \cdots \geq \rho_d$. Then:

$$\tilde{Q}_{i,S}(r, s) = e^{\alpha_r + \beta_s} (C_s - C_r) \mathbf{1}_{r<s} \quad (r \neq s),$$

$$\tilde{Q}_{i,S}(r, r) = e^{\beta_r - \bar{\beta}_r} \sigma \left(\bar{\beta}_r - \bar{\alpha}_r + \rho_r \right) \mathbf{1}_{r<d} + e^{\alpha_d - \bar{\alpha}_d} \mathbf{1}_{r=d},$$

where

$$\bar{\alpha}_k = \log \sum_{j=1}^{k} e^{\alpha_j}, \quad \bar{\beta}_k = \log \sum_{j=k+1}^{d} e^{\beta_j},$$

$$c_k = e^{-\bar{\beta}_k - \bar{\alpha}_k} \left(\sigma \left(\bar{\beta}_k - \bar{\alpha}_k + \rho_k \right) - \sigma \left(\bar{\beta}_k - \bar{\alpha}_k + \rho_{k+1} \right) \right),$$

$$C_t = \sum_{k=1}^{t-1} c_k, \quad \sigma(x) = \frac{1}{1 + e^{-x}}.$$

The derivation is provided in Appendix B. We write $\tilde{Q}_{i,S}$ instead of $Q_{i,S}$ due to the specific ordering of categories. The induced distribution of $v(S \cup i) \ominus v(S)$ is

$$\sum_{r<s} \tilde{Q}_{i,S}(r, s)\delta_{e_r - e_s} + \left(\sum_r \tilde{Q}_{i,S}(r, r) \right) \delta_0,$$

from which the off-diagonal entries of $\tilde{Q}_{i,S}(r, s)$ can be reconstructed.

Assume that $Q_{i,S}(r, s)$ are given for all S in a common ordering of the categories, in that $Q_{i,S}(r, s) = \tilde{Q}_{i,S}(\pi_S(r), \pi_S(s))$, where π_S is a permutation of $\{1, \ldots, d\}$ fulfilling the ordering condition used above. If

$$Q_i(r, s) = \mathbb{E}_{S \sim p^i} \left[Q_{i,S}(r, s) \right],$$

the distributions of Categorical values are given by

$$q_i = \sum_{r,s} Q_i(r, s)\delta_{e_r - e_s}.$$

The probability masses at each point $e_r - e_s \in E - E$ are interpretable as the probability (averaged over coalitions) that player i causes the payoff of v to flip from class s to class r.

We may define the following *query functional* on top of this distribution is

$$\ell_{mc} = \max_s \sum_{r \neq s} Q_i(r, s),$$

which quantifies the largest probability of any change in the output led by player i. It can be computed more efficiently as $\max_s Q_i(s) - Q_i(s, s)$, where the marginal distribution $Q_{S,i}(s)$ is given by

$$Q_{S,i}(s) = \mathbb{P}(v(S) = s) = e^{\beta_s - \bar{\beta}_0}.$$

B   Extended Derivation for Categorical Games

We provide a derivation of the expressions $\tilde{Q}_{i,S}(r, s)$. In this derivation, i and S are fixed, and we write $\mathcal{P}_{rs}$ for $\tilde{Q}_{i,S}(r, s)$. Let $d \geq 3$ be an integer, $[\alpha_j]$ and $[\beta_j]$ be sets of d real numbers. Above, $\alpha_j = \theta_{S \cup i, j}$ and $\beta_j = \theta_{S,j}$, but the derivation below does not make use of this. Also, let ε_j be d independent standard Gumbel variables, each of which has distribution function and density

$$F(\varepsilon) = \exp\left(e^{-\varepsilon}\right), \quad p(\varepsilon) = F(\varepsilon)' = \exp\left(-\varepsilon - e^{-\varepsilon}\right) = e^{-\varepsilon}F(\varepsilon).$$

Fix $r, s \in \{1, \ldots, d\}$, $r \neq s$. We would like to obtain an expression for the probability $\mathcal{P}_{rs}$ of

$$\arg\max_j \left(\alpha_j + \varepsilon_j\right) = r \quad \text{and} \quad \arg\max_j \left(\beta_j + \varepsilon_j\right) = s.$$

Define

$$\alpha_{jr} := \alpha_j - \alpha_r, \quad \beta_{js} := \beta_j - \beta_s.$$

The arg max equalities above can also be written as a set of $2d$ inequalities (2 of which are trivial):

$$\varepsilon_j \leq \varepsilon_r - \alpha_{jr}, \quad \varepsilon_j \leq \varepsilon_s - \beta_{js}, \quad j = 1, \ldots, d.$$

Then:

$$\mathcal{P}_{rs} = \mathbb{E}\left[\prod_j I_j\right], \quad I_j := \mathbf{1}_{\varepsilon_j \leq \min(\varepsilon_r - \alpha_{jr}, \varepsilon_s - \beta_{js})}.$$

Two of them are simple:

$$I_r = \mathbf{1}_{\varepsilon_r \leq \varepsilon_s - \beta_{rs}}, \quad I_s = \mathbf{1}_{\varepsilon_s \leq \varepsilon_r - \alpha_{sr}}, \quad I_r I_s = \mathbf{1}_{\alpha_s - \alpha_r \leq \varepsilon_r - \varepsilon_s \leq \beta_s - \beta_r}.$$

Denote

$$\gamma_j := \alpha_{jr} - \beta_{js} = \rho_j - (\alpha_r - \beta_s), \quad \rho_j := \alpha_j - \beta_j.$$

Note that γ_j depends on r, s, but ρ_j does not. If $j \neq r, s$, then

$$I_j = \mathbf{1}_{\varepsilon_j \leq \varepsilon_r - \alpha_{jr}} \mathbf{1}_{\varepsilon_r - \varepsilon_s \leq \gamma_j} + \mathbf{1}_{\varepsilon_j \leq \varepsilon_s - \beta_{js}} \mathbf{1}_{\varepsilon_r - \varepsilon_s \geq \gamma_j}.$$

If we exchange sum and product, we obtain an expression of $\mathcal{P}_{rs}$ as sum of 2^{d-2} terms. Each of these terms is an expectation over ε_r, ε_s, with the argument being the product of $d - 2$ terms $F(\varepsilon_r + a_j)$ or $F(\varepsilon_s + a_j)$ and a box indicator for $\varepsilon_r - \varepsilon_s$. In the sequel, we make this more concrete and show that at most $d - 1$ of these terms are nonzero.

    With a bit of hindsight, we assume that $\rho_1 \geq \rho_2 \geq \cdots \geq \rho_d$, which is obtained by reordering the categories. This implies that $[\gamma_j]$ is nonincreasing for all (r, s). Also, define the function $\pi(k) = k + \mathbf{1}_{r \leq k} + \mathbf{1}_{s-1 \leq k}$ from $\{1, \ldots, d - 2\}$ to $\{1, \ldots, d\} \setminus \{r, s\}$. We will argue in terms of a recursive computation over $k = 1, \ldots, d - 2$. Define

$$M_k(\varepsilon_r, \varepsilon_s) = \mathbb{E}\left[I_r I_s \prod_{1 \leq j \leq k} I_{\pi(j)} \,\middle|\, \varepsilon_r, \varepsilon_s\right], \quad k \geq 0,$$

so that $\mathcal{P}_{rs} = \mathbb{E}[M_{d-2}(\varepsilon_r, \varepsilon_s)]$. Each M_k can be written as sum of 2^k terms. Imagine a binary tree of depth $d - 1$, with layers indexed by $k = 0, 1, \ldots, d - 2$. Each node in this tree is annotated by a box indicator for $\varepsilon_r - \varepsilon_s$ and some information detailed below. We are interested in the 2^{d-2} leaf nodes of this tree.

B.1   Box Indicators. Which Terms Are Needed?

We begin with a recursive computation of the box indicators, noting that we can eliminate all nodes where the box is empty. Label the root node (at $k = 0$) by 1, its children
(at $k = 1$) by 10 (left), 11 (right), and so on, and define the box indicators as $\mathbf{1}_{l_1 \le \varepsilon_r - \varepsilon_s \le u_1}$,
and (l_{10}, u_{10}), (l_{11}, u_{11}) respectively. Then, $l_1 = \alpha_s - \alpha_r$, $u_1 = \beta_s - \beta_r$ defines the box for
the root. Here,

$$l_1 \ge u_1 \quad \Leftrightarrow \quad \rho_s \ge \rho_r.$$

Since $[\rho_j]$ is non-increasing, the root box is empty if $s < r$, so that $\mathcal{P}_{rs} = 0$ in this case.
In the sequel, we assume that $r < s$ and $\rho_r > \rho_s$, so that $l_1 < u_1$.
   If $\mathbf{n}$ is the label of a node at level $k - 1$ with box $(l_{\mathbf{n}}, u_{\mathbf{n}})$, then

$$l_{\mathbf{n}0} = l_{\mathbf{n}}, \quad u_{\mathbf{n}0} = \min(\gamma_{\pi(k)}, u_{\mathbf{n}}), \quad l_{\mathbf{n}1} = \max(\gamma_{\pi(k)}, l_{\mathbf{n}}), \quad u_{\mathbf{n}1} = u_{\mathbf{n}}.$$

Consider node 11 (right child of root). There are two cases. (1) $\gamma_{\pi(1)} < u_1$. Then, $l_{11} \ge$
$\gamma_{\pi(1)} \ge \gamma_{\pi(k)}$ for all $k \ge 1$, so all descendants must have the same $l = l_{11}$. If ever we step
to the left from here, $u = \min(\gamma_{\pi(k)}, u_1) \le \gamma_{\pi(k)} \le \gamma_{\pi(1)} \le l_{11}$, so the node is eliminated.
This means from 11, we only step to the right: 111, 1111, ..., with $l = \max(\gamma_{\pi(1)}, l_1)$,
$u = u_1$, so there is only one leaf node which is a descendant of 11. (2) $\gamma_{\pi(1)} \ge u_1$. Then,
$l_{11} \ge u_{11}$, so that 11 and all its descendants are eliminated.
   At node 10, we have $l_{10} = l_1$. If $\gamma_{\pi(1)} \le l_1$, the node is eliminated, so assume
$\gamma_{\pi(1)} > l_1$, and $u_{10} = \min(\gamma_{\pi(1)}, u_1)$. Consider its right child 101. We can repeat the
argument above. There is at most one leaf node below 101, with $l = \max(\gamma_{\pi(2)}, l_1)$ and
$u = u_{10} = \min(\gamma_{\pi(1)}, u_1)$.
   All in all, at most $d - 1$ leaf nodes are not eliminated, namely those with labels
$10\ldots01\ldots1$, and their boxes are $[\max(\gamma_{\pi(1)}, l_1), u_1]$, $[\max(\gamma_{\pi(2)}, l_1), \min(\gamma_{\pi(1)}, u_1)]$, ...,
$[\max(\gamma_{\pi(d-2)}, l_1), \min(\gamma_{\pi(d-3)}, u_1)]$, $[l_1, \min(\gamma_{\pi(d-2)}, u_1)]$.
   Recall that each node term is a product of $d - 2$ Gumbel CDFs times a box indicator.
What are these products for our $d - 1$ non-eliminated leaf nodes? The first is $F(\varepsilon_s -$
$\beta_{\pi(1)s}) \cdots F(\varepsilon_s - \beta_{\pi(d-2)s})$, the second is $F(\varepsilon_r - \alpha_{\pi(1)r})F(\varepsilon_s - \beta_{\pi(2)s}) \cdots F(\varepsilon_s - \beta_{\pi(d-2)s})$,
the third is $F(\varepsilon_r - \alpha_{\pi(1)r})F(\varepsilon_r - \alpha_{\pi(2)r})F(\varepsilon_s - \beta_{\pi(3)s}) \cdots F(\varepsilon_s - \beta_{\pi(d-2)s})$ and the last one
is $F(\varepsilon_r - \alpha_{\pi(1)r}) \cdots F(\varepsilon_r - \alpha_{\pi(d-2)r})$. Next, we derive expressions for the expectation of
these terms.

B.2   Analytical Expressions for Expectations

Consider $d - 2$ scalars $a_1, \ldots, a_{d-2}$ and $1 \le k \le d - 1$. We would like to compute

$$A = \mathbb{E}\left[\left(\prod_{j<k} F(\varepsilon_r + a_j)\right)\left(\prod_{j\ge k} F(\varepsilon_s + a_j)\right)\mathbf{1}_{l \le \varepsilon_r - \varepsilon_s \le u}\right]. \tag{4}$$

Denote

$$G(a_1, \ldots, a_t) := \mathbb{E}[F(\varepsilon_1 + a_1) \cdots F(\varepsilon_1 + a_t)].$$

We start with showing that

$$G(a_1, \ldots, a_t) = (1 + e^{-a_1} + \cdots + e^{-a_t})^{-1}.$$

Recall that $p(x) = F(x)' = e^{-x}F(x)$. If $\tilde{F}(x) = \prod_{j=1}^{t} F(x + a_j)$, then

$$\tilde{F}(x)' = \left(\sum_{j=1}^{t} e^{-a_j}\right) e^{-x}\tilde{F}(x).$$

Using integration by parts:

$$G(a_1, \ldots, a_t) = \int \tilde{F}(x)p(x)\,dx = 1 - \int \tilde{F}(x)'F(x)\,dx = 1 - \left(\sum_{j=1}^{t} e^{-a_j}\right) G(a_1, \ldots, a_t),$$

where we used that $F(x) = e^x p(x)$.

Next, define

$$g_1 = \log\left(1 + e^{-a_1} + \cdots + e^{-a_{k-1}}\right), \quad g_2 = \log\left(1 + e^{-a_k} + \cdots + e^{-a_{d-2}}\right).$$

We show that A in (4) can be written in terms of (g_1, g_2, l, u) only. Assume that $k > 1$ for now. Fix ε_s and do the expectation over ε_r. Note that $\mathbf{1}_{l \le \varepsilon_r - \varepsilon_s \le u} = \mathbf{1}_{\varepsilon_s + l \le \varepsilon_r \le \varepsilon_s + u}$. If $\tilde{F}(x) = \prod_{j<k} F(x + a_j)$, then

$$\tilde{F}(x)' = \left(\sum_{j<k} e^{-a_j}\right) e^{-x}\tilde{F}(x).$$

Using integration by parts:

$$B(\varepsilon_s) = \int_{\varepsilon_s+l}^{\varepsilon_s+u} \tilde{F}(x)p(x)\,dx = \left[\tilde{F}(x)F(x)\right]_{\varepsilon_s+l}^{\varepsilon_s+u} - B(\varepsilon_s)\sum_{j<k} e^{-a_j},$$

so that

$$B(\varepsilon_s) = e^{-g_1}\left[\tilde{F}(x)F(x)\right]_{\varepsilon_s+l}^{\varepsilon_s+u}$$

and

$$A = \mathbb{E}\left[B(\varepsilon_s)\prod_{j\ge k} F(\varepsilon_s + a_j)\right] = A_1 - A_2,$$

where

$$A_1 = e^{-g_1}\mathbb{E}\left[\left(\prod_{j<k} F(\varepsilon_s + u + a_j)\right)\left(\prod_{j\ge k} F(\varepsilon_s + a_j)\right)F(\varepsilon_s + u)\right]$$
$$= e^{-g_1}G(a_1 + u, a_2 + u, \ldots, a_{k-1} + u, a_k, \ldots, a_{d-2}, u)$$

and

$$A_2 = e^{-g_1}G(a_1 + l, a_2 + l, \ldots, a_{k-1} + l, a_k, \ldots, a_{d-2}, l).$$

Now,

$$-\log A_1 = g_1 - \log G(a_1 + u, a_2 + u, \ldots, a_{k-1} + u, a_k, \ldots, a_{d-2}, u)$$
$$= g_1 + \log\left(1 + \sum_{j<k} e^{-a_j-u} + \sum_{j\ge k} e^{-a_j} + e^{-u}\right) = g_1 + \log\left(e^{g_2} + e^{-u+g_1}\right)$$
$$= g_1 + g_2 + \log\left(1 + e^{g_1-g_2-u}\right)$$

and

$$-\log A_2 = g_1 + g_2 + \log\left(1 + e^{g_1-g_2-l}\right)$$

so that

$$A = A_1 - A_2 = e^{-(g_1+g_2)}\left(\sigma(g_2 - g_1 + u) - \sigma(g_2 - g_1 + l)\right), \quad \sigma(x) := \frac{1}{1 + e^{-x}}. \quad (5)$$

If $k = 1$, we can flip the roles of ε_r and ε_s by $g_1 \leftrightarrow g_2$, $l \to -u$, $u \to -l$, $k \to d - 1$, which gives

$$e^{-(g_1+g_2)}\left(\sigma(-(g_2 - g_1 + l)) - \sigma(-(g_2 - g_1 + u))\right)$$
$$= e^{-(g_1+g_2)}\left(\sigma(g_2 - g_1 + u) - \sigma(g_2 - g_1 + l)\right),$$

using $\sigma(-x) = 1 - \sigma(x)$, so the expression holds in this case as well.

B.3 Efficient Computation for All Pairs

Our $d - 1$ terms of interest can be indexed by $k = 1, \ldots, d - 1$. We can use the analytical expression just given with $a_j = -\alpha_{\pi(j)r}$ for $1 \leq j < k$ and $a_j = -\beta_{\pi(j)s}$ for $k \leq j \leq d - 2$. Define

$$g_1(k) = \log\left(1 + \sum\nolimits_{1 \leq j < k} e^{\alpha_{\pi(j)} - \alpha_r}\right), \quad g_2(k) = \log\left(1 + \sum\nolimits_{k \leq j \leq d-2} e^{\beta_{\pi(j)} - \beta_s}\right),$$

as well as

$$l(k) = \max(\gamma_{\pi(k)}, l_1), \quad u(k) = \min(\gamma_{\pi(k-1)}, u_1),$$

where we define $\pi(0) = 0$, $\pi(d - 1) = d + 1$, $\gamma_0 = +\infty$, and $\gamma_{d+1} = -\infty$. Note that

$$l(k) = \max(\rho_{\pi(k)} - \alpha_r + \beta_s, \alpha_s - \alpha_r) = \beta_s - \alpha_r + \max(\rho_{\pi(k)}, \rho_s),$$
$$u(k) = \min(\rho_{\pi(k-1)} - \alpha_r + \beta_s, \beta_s - \beta_r) = \beta_s - \alpha_r + \min(\rho_{\pi(k-1)}, \rho_r). \quad (6)$$

$\mathcal{P}_{rs}$ is obtained as sum of $A(g_1(k), g_2(k), l(k), u(k))$ for $k = 1, \ldots, d - 1$. In the sequel, we show how to compute these terms efficiently, for all pairs $r < s$.

Recall that $\gamma_j = \rho_j - (\alpha_r - \beta_s)$, $u_1 = \beta_s - \beta_r$, $l_1 = \alpha_s - \alpha_r$. Then:

$$l(k) < u(k) \quad \Leftrightarrow \quad \rho_{\pi(k)} < \rho_{\pi(k-1)} \wedge \rho_{\pi(k)} < \rho_r \wedge \rho_s < \rho_{\pi(k-1)}.$$

Recall that $\pi(k) = k + \mathbf{1}_{r \leq k} + \mathbf{1}_{s-1 \leq k}$. Define $K_1 = \{1, \ldots, r - 1\}$, $K_3 = \{s, \ldots, d - 1\}$, each of which can be empty. For $k \in K_1$, $\rho_{\pi(k)} = \rho_k \geq \rho_r$, so $l(k) \geq u(k)$. For $k \in K_3$, we have $\pi(k - 1) = k + 1 > s$, so that $\rho_s \geq \rho_{\pi(k-1)}$ and $l(k) \geq u(k)$. This means we only need to iterate over $k \in K_2 = \{r, \ldots, s - 2\}$ with $\pi(k) = k + 1$ and $k = s - 1$ with $\pi(k) = s + 1$ (the latter only if $s < d$).

As k runs in K_2, $\pi(k) = r + 1, \ldots, s - 1$, and if $s < d$ then $\pi(s - 1) = s + 1$. Now

$$g_1(k) = \log\left(1 + \sum\nolimits_{1 \leq j < k} e^{\alpha_{\pi(j)} - \alpha_r}\right) = \log \sum\nolimits_{1 \leq j \leq k} e^{\alpha_j - \alpha_r},$$

using that $e^{\alpha_r - \alpha_r} = 1$. For $g_2(k)$, if $k < s - 1$, then $\{\pi(j) \mid k \leq j \leq d - 2\} = \{k + 1, \ldots, d\} \setminus \{s\}$, and if $k = s - 1$, the same holds true (the set is empty if $s = d$). Using $e^{\beta_s - \beta_s} = 1$, we have

$$g_2(k) = \log \sum\nolimits_{k < j \leq d} e^{\beta_j - \beta_s}.$$

Define

$$\bar{\alpha}_k := \log \sum_{j=1}^{k} e^{\alpha_j}, \quad \bar{\beta}_k := \log \sum_{j=k+1}^{d} e^{\beta_j}, \quad k = 1, \ldots, d-1.$$

Then:

$$g_1(k) = \bar{\alpha}_k - \alpha_r, \quad g_2(k) = \bar{\beta}_k - \beta_s, \quad k = r, \ldots, s-1.$$

Finally, using $g_2(k) - g_1(k) = \bar{\beta}_k - \bar{\alpha}_k + \alpha_r - \beta_s$ and (6), we have

$$g_2(k) - g_1(k) + l(k) = \bar{\beta}_k - \bar{\alpha}_k + \max(\rho_{\pi(k)}, \rho_s),, \quad g_2(k) - g_1(k) + u(k) = \bar{\beta}_k - \bar{\alpha}_k + \min(\rho_{\pi(k-1)}, \rho_r).$$

Some extra derivation, distinguishing between (a) $r = s - 1$, (b) $r < s - 1 \wedge k \in K_2$, (c) $r < s - 1 \wedge k = s - 1$ shows that

$$\max(\rho_{\pi(k)}, \rho_s) = \rho_{k+1}, \quad \min(\rho_{\pi(k-1)}, \rho_r) = \rho_k, \quad k = r, \ldots, s-1.$$

Plugging this into (5):

$$A(k) = e^{\alpha_r + \beta_s} c_k, \quad c_k = e^{-\bar{\beta}_k - \bar{\alpha}_k} \left(\sigma\left(\bar{\beta}_k - \bar{\alpha}_k + \rho_k\right) - \sigma\left(\bar{\beta}_k - \bar{\alpha}_k + \rho_{k+1}\right) \right).$$

and $\mathcal{P}_{rs} = \sum_{k=r}^{s-1} A(k)$. Importantly, c_k does not depend on r, s. Therefore:

$$\mathcal{P}_{rs} = e^{\alpha_r + \beta_s}(C_s - C_r), \quad C_t = \sum_{k=1}^{t-1} c_k \quad (r < s); \quad \mathcal{P}_{rs} = 0 \quad (r > s). \tag{7}$$

The sequences $[\bar{\alpha}_k], [\bar{\beta}_k], [c_k], [C_k]$ can be computed in $O(d)$.

Finally, we also determine $\mathcal{P}_{rr}$, which is defined by the inequalities $\varepsilon_j \leq \varepsilon_1 - \max(\alpha_{jr}, \beta_{jr})$. A derivation like above (but simpler) gives:

$$\mathcal{P}_{rr} = \left(1 + \sum_{j \neq r} e^{\max(\alpha_{jr}, \beta_{jr})} \right)^{-1}.$$

Now, $\alpha_{jr} \geq \beta_{jr}$ iff $\rho_j \geq \rho_r$ iff $j < r$, so that

$$\mathcal{P}_{rr} = \left(1 + \sum_{j<r} e^{\alpha_j - \alpha_r} + \sum_{j>r} e^{\beta_j - \beta_r} \right)^{-1} = \left(e^{\bar{\alpha}_r - \alpha_r} + e^{\bar{\beta}_r - \beta_r} \right)^{-1}$$

$$= e^{\beta_r - \bar{\beta}_r} \sigma(\bar{\beta}_r - \bar{\alpha}_r + \rho_r), \quad (r < d),$$

$$\mathcal{P}_{dd} = e^{\alpha_d - \bar{\alpha}_d}.$$

C   Related Work in Cooperative Game Theory

The Shapley value of simple game has a probabilistic interpretation (Peleg and Sudhölter, 2007, pag. 168) however simple games are not Categorical games. An and-or axiom substitute the linear axioms in simple games (Weber, 1988), here we address probabilisitc combinations. Stochastic games are typically intended as multi-stage games where the transition between stages is stochastic Shapley (1953b), Petrosjan (2006) and not the intrinsic payoffs. Static cooperative games with stochastic output have been considered from the perspective of coalition formation and considering notions of players' utility (e.g. Suijs et al., 1999) or studying two stages setups – before and after the realisation of the payoff (e.g. Granot, 1977), and from an optimization perspective (Sun et al., 2022). To the best of our knowledge, our settings and constructions have not been studied before.

References

Aas, K., Jullum, M., Løland, A.: Explaining individual predictions when features are dependent: more accurate approximations to shapley values. Artif. Intell. **298**, 103502 (2021)

Bhatt, U., et al.: Explainable machine learning in deployment. In: Proceedings of the 2020 Conference on Fairness, Accountability, and Transparency, pp. 648–657 (2020)

Chen, J., Song, L., Wainwright, M.J., Jordan, M.I.: L-shapley and C-shapley: efficient model interpretation for structured data. arXiv preprint arXiv:1808.02610 (2018)

Frye, C., Rowat, C., Feige, I.: Asymmetric shapley values: incorporating causal knowledge into model-agnostic explainability. Adv. Neural. Inf. Process. Syst. **33**, 1229–1239 (2020)

Granot, D.: Cooperative games in stochastic characteristic function form. Manage. Sci. **23**(6), 621–630 (1977)

He, K., Zhang, X., Ren, S., Sun, J.: Deep residual learning for image recognition (2015)

Heskes, T., Sijben, E., Bucur, I.G., Claassen, T.: Causal shapley values: exploiting causal knowledge to explain individual predictions of complex models. Adv. Neural Inf. Process. Syst. **33**, 4778–4789 (2020)

Kermany, D.S., et al.: Identifying medical diagnoses and treatable diseases by image-based deep learning. Cell **172**(5) (2018). https://doi.org/10.1016/j.cell.2018.02.010

Kim, Y., Denton, C., Hoang, L., Rush, A.M.: Structured attention networks. arXiv preprint arXiv:1702.00887 (2017)

Kumar, I.E., Venkatasubramanian, S., Scheidegger, C., Friedler, S.: Problems with shapley-value-based explanations as feature importance measures. In: International Conference on Machine Learning, pp. 5491–5500. PMLR (2020)

Lundberg, S.M., Lee, S.I.: A unified approach to interpreting model predictions. Adv. Neural Inf. Process. Syst. **30** (2017)

Miller, T.: Explanation in artificial intelligence: Insights from the social sciences. Artif. Intell. **267**, 1–38 (2019)

Mittelstadt, B., Russell, C., Wachter, S.: Explaining explanations in AI. In: Proceedings of the Conference on Fairness, Accountability, and Transparency, pp. 279–288 (2019)

Papandreou, G., Yuille, A.L.: Perturb-and-map random fields: using discrete optimization to learn and sample from energy models. In: 2011 International Conference on Computer Vision, pp. 193–200. IEEE (2011)

Peleg, B., Sudhölter, P.: Introduction to the Theory of Cooperative Games, vol. 34. Springer, Heidelberg (2007). https://doi.org/10.1007/978-3-540-72945-7

Petrosjan, L.A.: Cooperative stochastic games. In: Haurie, A., Muto, S., Petrosjan, L.A., Raghavan, T.E.S. (eds.) Advances in Dynamic Games. Annals of the International Society of Dynamic Games, vol. 8, pp. 139–145. Springer, Boston (2006). https://doi.org/10.1007/0-8176-4501-2_7

Roth, A.E.: The Shapley Value: Essays in Honor of Lloyd S. Shapley. Cambridge University Press, Cambridge (1988)

Shapley, L.: A Value for n-Person Games, p. 343 (1953a). Artin, E., Morse, M. (eds.)

Shapley, L.S.: Stochastic games. Proc. Natl. Acad. Sci. **39**(10), 1095–1100 (1953b)

Suijs, J., Borm, P., De Waegenaere, A., Tijs, S.: Cooperative games with stochastic payoffs. Eur. J. Oper. Res. **113**(1), 193–205 (1999)

Sun, P., Hou, D., Sun, H.: Optimization implementation of solution concepts for cooperative games with stochastic payoffs. Theor. Decis. **93**(4), 691–724 (2022)

Sundararajan, M., Najmi, A.: The many Shapley values for model explanation. In: International Conference on Machine Learning, pp. 9269–9278. PMLR (2020)

Vaswani, A., et al.: Attention is all you need. Adv. Neural Inf. Process. Syst. **30** (2017)

Weber, R.J.: Probabilistic Values for Games. The Shapley Value. Essays in Honor of Lloyd S. Shapley, pp. 101–119 (1988)

Privacy-Preserving Machine Learning for Healthcare: Open Challenges and Future Perspectives

Alejandro Guerra-Manzanares$^{(\boxtimes)}$, L. Julian Lechuga Lopez ,
Michail Maniatakos , and Farah E. Shamout

Department of Computer Engineering, New York University Abu Dhabi,
Abu Dhabi, UAE
{ag9454,1jl5178,mm6446,fs999}@nyu.edu

Abstract. Machine Learning (ML) has recently shown tremendous success in modeling various healthcare prediction tasks, ranging from disease diagnosis and prognosis to patient treatment. Due to the sensitive nature of medical data, privacy must be considered along the entire ML pipeline, from model training to inference. In this paper, we conduct a review of recent literature concerning Privacy-Preserving Machine Learning (PPML) for healthcare. We primarily focus on privacy-preserving training and inference-as-a-service, and perform a comprehensive review of existing trends, identify challenges, and discuss opportunities for future research directions. The aim of this review is to guide the development of private and efficient ML models in healthcare, with the prospects of translating research efforts into real-world settings.

Keywords: privacy-preserving · machine learning · healthcare

1 Introduction

Machine Learning (ML) and Deep Learning (DL) have shown great promise in many domains, leveraging the use of large datasets. Some notable contributions include *AlphaFold* [25] for the prediction of protein structures and *Transformers* [59] for natural language processing. Healthcare is one of the domains in which ML is expected to provide substantial improvements in the delivery of patient care worldwide [64]. Given the rapid growth in the number of models over the last couple of years [23,28,38,50], healthcare applications deserve special consideration considering the sensitive nature of the data that is required to train the models and the safety-critical nature of medical decision-making.

In this regard, real-world implementation of such models is still hampered by ethical and legal constraints. Legal frameworks have been developed and enforced to guarantee the transparency and privacy of ML-based healthcare solutions, such as the *Health Insurance Portability and Accountability Act (HIPAA)* in

---

A. Guerra-Manzanares and L. J. L. Lopez—Equal contributions.

H. Chen and L. Luo (Eds.): TML4H 2023, LNCS 13932, pp. 25–40, 2023.
https://doi.org/10.1007/978-3-031-39539-0_3

the United States [17] and the *General Data Protection Regulation (GDPR)* in Europe [61]. Therefore, there is a crucial need for Privacy-Preserving Machine Learning (PPML) in healthcare to enable the implementation of trustworthy systems in the future. The main goal of this review is to provide a comprehensive overview of state-of-the-art PPML in healthcare and encourage the development of new methodologies that tackle specific challenges relevant to the nature of the domain.

Motivation. There exist several related literature reviews that focus on a specific subset of PPML for healthcare. Several highlight recent advancements in federated learning [3,24,40,68], cryptographic techniques [73], or security aspects of ML models, such as adversarial attacks [33]. Existing review articles cover a wide range of applications related to health and input data modalities, ranging from IoT sensors to medical images [48]. Compared to existing work, our review has three main contributions with the intent of bridging between research pertaining to ML for healthcare and cybersecurity. First, we distinguish between PPML for training and inference, i.e., *ML-as-a-service*. Second, we focus on state-of-the-art (SOTA) literature published in the last three years, considering the high proliferation of ML in healthcare and recent methodological advancements in ML and DL (e.g., network architectures, model pre-training, etc.). Third, we consider studies that develop or apply methodologies using two popular modalities based on publicly available datasets and state-of-the-art in ML for healthcare, namely medical images and data extracted from Electronic Health Records (EHR) [28]. Despite the use of other input modalities in medical applications, such as video [44] or text [56], our review exclusively focuses on medical images and EHR as they are the most prevalent input modalities in diagnostic and prognostic settings [53]. Lastly, although we acknowledge the importance of security for ML models, it is out of the scope of this paper since we primarily focus on privacy.

To this end, we review papers that meet the following inclusion criteria:

1. We include recently published work i.e., publication year $\geq$ 2020.
2. We include articles that focus on the application or development of PPML either for model training and/or inference, including but not restricted to homomorphic encryption, differential privacy, federated learning, and multi-party secure computation.
3. We include articles that consider clinical tasks involving medical images and/or EHR data.

In Sect. 2, we provide background knowledge about concepts and terminology concerning PPML. In Sect. 3, we provide an overview of the state-of-the-art pertaining to PPML for training (Sect. 3.1) and for inference (Sect. 3.2). Later in Sect. 4, we discuss open challenges and derive future directions. Finally, we provide concluding remarks in Sect. 5.

2 Privacy-Preserving Machine Learning: Background and Terminology

2.1 Federated Learning

Since medical data is highly sensitive, data sharing is difficult, and subject to ethical restrictions and legal constraints if at all possible. Federated learning (FL) [37] aims to overcome the challenges of data sharing by enabling collaborative training, which does not require that the involved parties share their training data. Therefore, the data remains private to each local node within the FL network, such that only the model updates are shared and integrated in a centralized model.

Federated averaging [37] is the most common form of FL. In this setting, a centralized server is connected to N entities, which have their own training data. The central server orchestrates the collaborative training process as follows: (1) the initial model is distributed amongst all entities, (2) each entity performs a training iteration on their local model using their own training data, typically one epoch, and shares its resulting model parameters with the central server, (3) the server averages the model parameters shared by all entities and distributes the resulting (averaged) model amongst all entities, and (4) steps (2) and (3) are repeated sequentially until a performance threshold or a specific number of training iterations is achieved. FL has proven to be very efficient in training models with strong performance, while avoiding the need for data sharing [37]. However, FL might be vulnerable to privacy issues such as reconstruction attacks [34], thus requiring that it is combined with other privacy-preserving methods to ensure robust privacy guarantees [40].

2.2 Differential Privacy

Differential Privacy (DP) has its origins in statistical analysis of databases. Its main aim is to address the paradox of learning nothing about specific individuals, while learning useful information about the general population [12]. In the FL context, it is usually incorporated in the form of additive noise to model updates, either artificially or using a differentiable private optimizer, prior to transferring the updates from the entities to the central server [1]. The amount of artificial noise added is directly proportional to the degree of privacy desired (i.e., privacy budget) [77]. DP can successfully make privacy attacks fail, such as reconstruction attacks, as the added noise hinders the inference of actual knowledge about the training data by the attacker. However, adding too much noise (i.e., high privacy budget) can hamper learning and negatively impact the model accuracy [7].

2.3 Homomorphic Encryption

In mathematics, the term *homomorphic* refers to the transformation of a given set into another while preserving the relation between the elements in both sets. Thus, Homomorphic Encryption (HE) refers to the conversion of plaintext into ciphertext while preserving the structure of the data. Consequently, specific operations applied to the ciphertext will provide the same results as if they were applied to the plaintext but without compromising the encryption [2]. That is, the plaintext data is never accessed nor decrypted as the operations are directly applied to the encrypted data. The result of the transformations on the ciphertext can only be decrypted back to plaintext by the encryption key owner.

Despite the benefit of provable privacy guarantees, the range of operations available in HE is restricted to addition and multiplication i.e., *fully* homomorphic encryption. This limits the set and number of transformations applicable to the data and requires the use of approximations for more complex operations (e.g., HE-ReLU is the polynomial approximation of the ReLU function [72]). This also significantly increases the computational time needed to process encrypted text compared to plaintext by several orders of magnitude [47].

2.4   Secure Multi-party Computation

Secure Multi-Party Computation (SMPC) [15] provides a framework in which two or more parties jointly compute a public function with their data while keeping the inputs private and hidden from other parties using cryptographic protocols. Most protocols used for SMPC with more than two parties are based on Secret Sharing (SS). In SS, a portion of the secret input is shared among a number of other parties. Most ML methods use Shamir's SS and additive SS [54]. Although these methods are considered information-theoretic secure cryptosystems, recent studies show that leakage of global data properties can occur in some scenarios [76]. While both FL and SMPC rely on collaborative training via knowledge sharing and keep the end-point data private, their implementation differs significantly. SMPC involves cryptography and can be used for training and inference, whereas FL does not involve cryptography nor provides strong privacy guarantees, and is only used for model training.

3   Overview of State-of-the-Art

Following the inclusion criteria described in Sect. 1, we summarize existing work on PPML for healthcare based on whether the work focuses on model training (Table 1) or model inference (Table 2). For each study (row) we describe several attributes. *Use case* provides a succinct summary of the objective of the study. *Model* reports the ML or DL architecture that was employed to model the task. *Medical datasets* summarizes the datasets that were used for model training and evaluation. Additionally, we use the * symbol to indicate the use of a private dataset. *ML task* describes the nature of the prediction task (e.g., binary or multi-class classification). *Input modality* reports the nature of the model's input data, which could either be I for medical images or E for EHR data. In the *Validation* column, we report whether the trained model was internally and/or externally evaluated, with ✔ indicating the use of internal validation i.e., test set from the same distribution of the training data, and ✔✔ indicating the assessment of the generalization of the model on an external test dataset. Lastly, *Metrics* lists the evaluation metrics used to describe the performance of the proposed model (Table 1).

3.1 Privacy-Preserving Training for Healthcare

Table 1. Summary of PPML in healthcare for model training. We summarize studies that focus on developing PPML in the context of model training. We group them based on the methodology considered, i.e. federated learning, homomorphic encryption, and differential privacy.

Reference	Use case	Model	Medical dataset/s	ML task	Input modality	Validation	Metrics
			FEDERATED LEARNING				
[11]	COVID-19 Computed Tomography (CT) analysis	RetinaNet	Multi-institution lung CT data*	Object detection	I	✔✔	mAP, Specifity, Recall, AUROC
[14]	Cardiovascular admission after lung cancer treatment	Logistic regression	Multi-institution lung CT data*	Risk prediction	I+E	✔	AUROC, C-index
[31]	FL benchmarking and reliability in healthcare	Neural Network, LSTM, CNN	MIMIC-III, PhysioNet ECG	Mortality prediction, Multi-class classification	I+E	✔	AUROC, AUPRC, F1-score
[71]	FL benchmarking and monetary cost in healthcare	Transformer, EfficientNet-B0, ResNet-NC-SE	eICU, ISIC19, HAM10000, PhysioNet ECG	Mortality prediction, Length of stay, Discharge time, Acuity prediction	I+E	✔	AUROC, AUPRC
[51]	FL benchmarking vs. centralized learning in healthcare	Logistic regression, Neural Network, Generalized linear model	UCI Heart failure, MIMIC-III, Malignancy in SARS-CoV-2 infection	Risk prediction	E	✔	AUROC
[35]	COVID-19 detection	DenseNet	Multi-institution COVID-19 X-ray*	Binary classification	I	✔✔	AUROC, AUPRC
[67]	Coronary artery calcification (CAC) forecast	Random Forest	CAC risk factors*	Risk prediction	E	✔	Recall, Specificity
[62]	Cancer inference via gene expression	Gradient Boosting Decision Tree	iDASH 2020	Multi-class classification	E	✔	Accuracy, AUC, Recall, Precision, F1-score
[20]	Diabetic kidney risk prediction	Logistic regression, MLP	CERNER Health Facts	Risk prediction	E	✔	F1-score
[9]	Lung cancer post-treatment 2-year survival	Logistic regression	Multi-institution lung cancer EHR*	Mortality prediction	E	✔	RMSE, Accuracy, AUROC
[45]	COVID-19 detection	Transformer with DenseNet, TransUNet and RetinaNet	Multi-institution COVID-19 X-ray (public and private datasets)	Multi-task: classification, segmentation, object detection	I	✔✔	AUC, mAP, Dice coefficient
[69]	Multiple medical prediction tasks	Self-supervised vision transformer	COVID-19 X-ray, Kaggle Diabetic Retinopathy, Dermatology ISIC	Binary/multi-class classification, Object detection	I	✔✔	Accuracy, F1-score
			HOMOMORPHIC ENCRYPTION				
[5]	COVID-19 detection	MobileNet-V2	COVID-19 X-ray	Multi-class classification	I	✔	Accuracy, Recall, Precision, F1-score
[36]	Heart and thyroid disease classification	XGBoost	UCI Heart Disease, Kaggle Hypothyroid	Binary classification	E	✔	Accuracy
[46]	Intensive Care Unit patient outcome	LSTM	MIMIC-III	Binary classification	E	✔	Recall, AUROC, Precision
[6]	Dermatology diagnostics	SVM	UCI Dermatology	Multi-class classification	E	✔	Accuracy
[4]	COVID-19 detection	AlexNet, SqueezeNet	COVID-19 X-ray, COVID-19 CT	Multi-class classification	I	✔	Accuracy, F1-score

(continued)

Table 1 Continued. Summary of PPML in healthcare for model training.
We summarize here studies that use a combination of federated learning and other
privacy-preserving techniques, blockchain, Secure Multi-Party Computation (SMPC),
image encryption, and image modification.

Reference	Use case	Model	Medical dataset/s	ML task	Input modality	Validation	Metrics
DIFFERENTIAL PRIVACY							
[77]	Thoracic pathology detection	DenseNet-121	CheXpert	Multi-class classification	I+E	✔	AUROC, Accuracy
[7]	COVID-19 detection	EfficientNet-B2	COVID-19 X-ray	Binary classification	I	✔	Accuracy
[57]	Multiple medical prediction tasks	CNN, DenseNet-121, Logistic regression, GRU-D	MNIST NIH Chest X-ray, MIMIC-III	Binary, Multi-class classification	I+E	✔	AUROC
FEDERATED LEARNING + DIFFERENTIAL PRIVACY							
[21]	Cardiomyopathy risk prediction	Random Forest, Naive Bayes	iDASH 2021, Breast Cancer TCGA	Risk prediction	E	✔	AUROC
[29]	In-hospital mortality prediction	CNN	Premier Healthcare Database*	Mortality prediction	E	✔	AUROC, Overhead
[8]	COVID-19 patient triage	ResNet-34 DeepCrossNet	Multi-institution chest x-ray and EHR*	Risk prediction	I+E	✔✔	AUROC, Recall, Specificity
BLOCKCHAIN							
[74]	Distributed training	ResNet-18	NSCLC-Radiomics	Binary classification	I	✔✔	AUROC
[63]	Disease classification	Neural Network	Blood transcriptomes*	Binary classification	E	✔	Accuracy
FEDERATED LEARNING + HOMOMORPHIC ENCRYPTION							
[65]	COVID-19 detection	CNN	COVID-19 X-ray	Binary classification	I	✔	Accuracy, Recall Precision, F1 score, Execution time
FEDERATED LEARNING + HOMOMORPHIC ENCRYPTION + SMPC							
[75]	Skin cancer classification	CNN	HAM10000	Multi-class classification	I	✔	Accuracy, Overhead
SMPC							
[18]	Tumor detection	Logistic regression	iDASH 2019	Binary classification	E	✔	Accuracy, Overhead
IMAGE ENCRYPTION							
[19]	Brain tumor, COVID-19	DenseNet-121, XceptionNet	MRI Brain Tumor, COVID-19 X-ray	Multi-class classification	I	✔	F1-score
IMAGE MODIFICATION							
[39]	Glaucoma recognition	VGAN-based CNN	Warsaw-BioBase Disease-Iris v2.1	Binary classification	I	✔	F1-score, Accuracy

As observed in Table 1, the most commonly used privacy-preserving approach
for model training is FL, either independently or in combination with DP. DP
is added to increase the privacy of the FL training updates i.e., adding noise to
the shared weights, thus making the system more robust to privacy threats, such
as reconstruction attacks by an external actor intercepting the communication
channel or an *honest-but-curious* central server [40].

The second most commonly investigated approach for private training is HE, which leverages encryption schemes to provide privacy with provable mathematical guarantees. However, as described in the previous section, training ML models on encrypted data significantly increases the computational complexity and the processing overhead by several orders of magnitude [65,75]. It also adds noise to the training process due to the approximations of activation functions, especially in large models.

The third most common approach is standalone DP, which is less computationally demanding and provides strong privacy guarantees. However, the increase in privacy guarantees is negatively correlated with model accuracy, as it is associated with an increase in the quantity of noise applied. Therefore, the trade-off between privacy (i.e., privacy budget) and model accuracy is a relevant factor to take into account for the inclusion of DP in any ML solution. There are other PPML approaches for model training that have been evaluated in related work, including the addition of a blockchain ledger to avoid the centralization of training (i.e., fully distributed learning), image modification to increase data privacy in the context of model explainability, and SMPC as an alternative encryption scheme to HE.

Most of the reviewed studies use a single source of input data i.e., image or EHR and only one medical dataset. Although some studies train their models on several datasets, including popular computer vision benchmarks, the vast majority restrict their evaluation to one input modality from the same dataset.

This limits the generalization of the results and neglects the potential improvement in predictive performance that could result from combining different data sources in multi-modal learning settings [49]. Furthermore, most studies perform internal validation, such that the test sets are from the same distribution as the training dataset. This is generally a common challenge in healthcare applications considering distribution shifts across different hospitals, for example due to differences in patient demographics. Finally, most existing work focuses on convolutional neural networks to handle computer vision tasks. However, validation schemes and metrics reported are not consistent, making the comparison among them very difficult. Due to these reasons and the lack of medical benchmark datasets, a fair comparison of the approaches is difficult, and therefore we do not assess performance metrics results in this review and defer it to future work.

3.2   Privacy-Preserving Inference for Healthcare

We now focus on the literature employing PPML methods for inference, as summarized in Table 2. We frame PPML for inference as providing private *machine-learning-as-a-service* (MLaaS) or *inference-as-a-service* (IaaS) [32]. In this scenario, a model with strong performance is controlled by a single party (i.e., model owner), and other external parties (i.e., clients) would like the model to perform inference on their own data. The external parties can share data samples with the model owner and their predictions are sent back. Due to legal and/or ethical constraints related to privacy, clients cannot disclose their data with the model owner, thus requiring the use of PPML to maintain the privacy of the data they wish to share.

Table 2. Summary of PPML in healthcare for model inference. We summarize studies that focus on developing PPML in the context of model inference. We group them based on the methodology considered, i.e. homomorphic encryption, combination of federated learning and Secure Multi-Party Computation (SMPC), differential privacy and SMPC, federated learning with blockchain and SMPC, and finally federated learning with differential privacy and homomorphic encryption.

Reference	Use case	Model	Medical dataset/s	ML task	Input modality	Validation	Metrics
HOMOMORPHIC ENCRYPTION							
[72]	Breast and cervical cancer classification	Convolutional LSTM	Cervigram Image, BreaKHis	Binary, Multi-class classification	I	✔	AUROC
[58]	Breast cancer classification	Neural Network, SVM	UCI IRIS, UCI Breast Cancer	Binary, Multi-class classification	E	✔	Accuracy, Privacy budget, Overhead
[52]	Cancer inference via gene expression	SVM, Logistic regression, Neural Network	iDASH 2020	Multi-class classification	E	✔✔	Accuracy, AUROC
[60]	Coronary angiography view classification	CNN	X-ray coronary angiography*	Binary, Multi-class classification	I	✔	Accuracy
FEDERATED LEARNING + SMPC							
[78], [26]+	Paediatric chest X-ray classification	ResNet-18	Chest X-ray*	Multi-class classification	I	✔✔	AUROC, Latency
DIFFERENTIAL PRIVACY + SMPC							
[54]	Pneumonia detection	CNN, VGG-16	Kaggle X-ray Pneumonia	Binary classification	I	✔	Accuracy
[22]	Accuracy-privacy trade-off analysis	Neural Network	Kaggle IDC, MIMIC-III	Binary, Multi-class classification	I	✔	Accuracy, Recall Precision, Privacy
FEDERATED LEARNING + BLOCKCHAIN + SMPC							
[27]	Multiple medical image datasets classification	CNN	MedMNIST (CXR, Breast, Hand, ChestCT, Abdomen, HeadCT)	Multi-class classification	I	✔	Accuracy
FEDERATED LEARNING + DIFFERENTIAL PRIVACY + HOMOMORPHIC ENCRYPTION							
[16]	Multiple medical image datasets classification	CNN	MedMNIST (Pneumonia Breast, Retina, Blood)	Multi-class classification	I	✔	Accuracy, Execution time, Bandwidth

+ [26] is an extension of [78].

Compared to the number of studies addressing PPML for training, a relatively fewer number have explored PPML for inference. Most studies within the theme of PPML for inference, focus on the deployment of the trained model as a service and its use by third parties. The most common approach for delivering PPML IaaS is HE, which ensures with provable mathematical guarantees that neither the model owner nor any intermediate party are able to inspect the original data nor the detection result i.e., both are encrypted and can only be decrypted by the data owner. Another common approach is SMPC, which also leverages encryption schemes, being used in combination with other privacy-preserving collaborative approaches such as FL, DP and blockchain.

Similar to PPML for training, most studies here use a single source of input data (i.e., images in most cases), neglecting many other diverse medical modalities of varying characteristics. The lack of use of benchmark medical datasets and inconsistent validation schemes and metrics hinders the generalization of the proposed approaches.

3.3   Open Challenges

There is No *One-Size-Fits-All* PPML Approach for Model Training or Inference by Design. We observe that previous work pick and choose PPML approaches based on the intended clinical use case. Currently, there is no consensus on what different "privacy models" look like in healthcare. Since the methodology depends on the use case, we also observe a clear trade-off between privacy and accuracy, based on the availability of computational resources. For instance, standalone FL is computationally faster than HE, but it does not provide strong privacy guarantees. On the other hand, HE and DP can provide strong privacy guarantees but they add noise to the model both for private training and private inference resulting in less accurate solutions. For HE, this is especially critical for model training where successive layers of approximations are needed to perform operations that are not supported, such as *softmax*, or that are computationally inefficient, such as max pooling. In general, encryption-based options are provably secure but computationally inefficient, since they increase the processing overhead of training using cyphertext compared to plaintext data.

Additionally, the availability of computational resources is a decisive factor in choosing a particular PPML methodology. For instance, in the HE scenario, sending data over a communication channel does not require infrastructure for model training but still requires handling the encryption/decryption process appropriately. In the FL context, it requires that the entity has allocated resources for model training.

The Centralization of Model Training in FL Poses an Additional Security Threat. Relying on a single central server entails a single point of failure that is highly susceptible to security attacks such as Denial-of-Service. Although blockchain has been proposed to achieve fully distributed training and mitigate this threat, it increases the complexity of the information technology infrastructure significantly, requiring dedicated resources for the implementation of the distributed ledger and modeling framework.

Most Existing Work Use a Single Dataset and Do Not Conduct External Validation, thus Arising Concerns About the Generalization of the Results. We observe that existing work focus on a limited set of medical datasets. Additionally, some work only evaluate their solutions on computer vision benchmark datasets (e.g., *MNIST* or *CIFAR-10*) inferring that good performance on these datasets will provide similar results on medical image data [13,43]. However, this assumption is not empirically supported by work that uses both medical and non-medical datasets [16,22,57,60,77] and must, therefore, be avoided.

MLaaS for Healthcare has not been Explored Thoroughly. As demonstrated by the limited literature on this topic, we observe that the literature is highly skewed towards PPML for training. Considering disparities in technical capabilities and expertise, information technology resources, and availability of data across medical institutions, the case in which an entity does not have enough resources to perform model training independently is highly likely. Thus, the usage of third-party models as inference systems that can run on proprietary data is a prominent scenario that has not been thoroughly explored and should be considered in future research. MLaaS can provide access to models with strong performance, enabling full preservation of data privacy using PPML methods. This makes it a more efficient solution for small-scale or low-resource medical entities to access and leverage third-party knowledge.

4 Future Research Directions

4.1 Comprehensive Evaluation on Diverse Medical Datasets

For the sake of comparison and generalization of results, studies should complement their internal dataset evaluation with additional extensive evaluation on benchmark medical datasets. This is due to the fact that most of the existing work use a single dataset and do not perform external validation. The number of studies that use external datasets for validation is marginal. Only 9 out of the 40 studies considered validated their results with an independent test set. This hampers model generalization and hinders performance comparison among approaches built for the same medical task. For benchmarking, we suggest *MedMNIST* [70], which contains curated datasets for different medical tasks and modalities. Therefore, similar to *MNIST* or *CIFAR-10* for computer vision models, this medical dataset could be employed as a common benchmark for medical applications.

4.2 Multi-modal Models

Current advances in ML for healthcare are moving towards multi-modal learning, where several sources of information are combined to improve performance [49,55]. This approach not only tends to provide better performance but also ensures a comprehensive understanding of the different physiological variables involved in studying and modeling the development of human biology and pathology. As observed in Sects. 3.1 and 3.2, most work is restricted to a single modality. To develop robust and strong ML models, the use of different data sources to develop multi-modal systems is paramount. Notwithstanding that, the use of more clinical data entails more privacy concerns (e.g., individuals may be identified using correlated data) and requires more training resources due to increased model complexity. Therefore, additional privacy and computational constraints must be considered in the design of these algorithms.

4.3 Machine Learning as a Service (MLaaS)

The deployment of PPML within MLaaS is a very promising opportunity to access strong proprietary models by less resourceful institutions. Indeed, one of the main objectives of ML in healthcare is to develop efficient and scalable solutions that improve healthcare delivery. In addition to lack of resources, the deployment of these systems in medical settings can also be highly challenging [30,66]. The development of MLaaS is significantly less investigated than PPML for model training. Therefore, further research on this topic is required to provide secure, private and efficient data sharing between third-party model providers and client institutions. Reducing obstacles for clinical institutions to access powerful inference systems could lead to a major improvement in healthcare delivery across regions, bypassing physical barriers. It can also lead to an increase in the confidence and widespread adoption of ML in healthcare. It is important to note that the success of MLaaS is dependent on improvements in model generalizability and fairness in external datasets.

4.4 Integration of SOTA and Advances in Deep Learning

Future work should also investigate the integration of recent advances in DL and ML models in healthcare, considering that most of the current PPML work focuses on convolutional neural networks. For instance, the *Transformer* architecture and its variants [10], which are considered the current SOTA for many computer vision or natural language processing tasks, are only adopted by [45,71], and [69] in the current related literature. Adopting SOTA architectures can take advantage of the latest advances in research, both in terms of optimizing hardware and software, to maintain performance improvements in clinical prediction tasks.

4.5 Global and Local Explainability

Transparency and model explainability are essential for trustworthy artificial intelligence [42]. However, PPML methods, such as data encryption or noise addition, hinder global model and local prediction explainability. The collision between two key principles for trustworthy artificial intelligence, secure and PPML [41] and explainability, highlights an important research problem that is currently under-investigated. Only [39] attempt to address this problem, which should encourage future work in this research direction.

5   Conclusion

In this paper, we introduce and summarize recent literature concerning PPML for model training and inference in the healthcare domain. We highlight trends, challenges and promising future research directions. In conclusion, we recognize the lack of consensus when it comes to defining the requirements of privacy-preserving frameworks in healthcare. This requires collaboration between machine learning scientists, healthcare practitioners, and privacy and security experts. From the perspective of advancing ML approaches, we encourage researchers to perform comprehensive evaluation of proposed algorithms on diverse medical datasets to increase generalization, to investigate the constraints of PPML in multi-modal learning settings, to further consider the promise of

MLaaS in healthcare as a catalyst for improved healthcare delivery, and to adopt state-of-the-art advances in deep learning architectures to enhance model performance. Our suggestions aim to address research gaps and guide future research in PPML to facilitate the future adoption of trustworthy and private ML for healthcare.

Acknowledgements. This work was supported by the NYUAD Center for Interacting Urban Networks (CITIES), funded by Tamkeen under the NYUAD Research Institute Award CG001, and the Center for Cyber Security (CCS), funded by Tamkeen under NYUAD RRC Grant No. G1104.

References

1. Abadi, M., et al.: Deep learning with differential privacy. In: Proceedings of the 2016 ACM SIGSAC Conference on Computer and Communications Security, pp. 308–318 (2016)
2. Acar, A., Aksu, H., Uluagac, A.S., Conti, M.: A survey on homomorphic encryption schemes: theory and implementation. ACM Comput. Surv. (Csur) **51**(4), 1–35 (2018)
3. Ali, M., Naeem, F., Tariq, M., Kaddoum, G.: Federated learning for privacy preservation in smart healthcare systems: a comprehensive survey. IEEE J. Biomed. Health Inform. (2022)
4. Baruch, M., Drucker, N., Greenberg, L., Moshkowich, G.: A methodology for training homomorphic encryption friendly neural networks. In: Zhou, J., et al. (eds.) ACNS 2022. LNCS, vol. 13285, pp. 536–553. Springer, Cham (2022). https://doi.org/10.1007/978-3-031-16815-4_29
5. Boulila, W., Ammar, A., Benjdira, B., Koubaa, A.: Securing the classification of COVID-19 in chest x-ray images: a privacy-preserving deep learning approach. In: 2022 2nd International Conference of Smart Systems and Emerging Technologies (SMARTTECH), pp. 220–225. IEEE (2022)
6. Chen, Y., Mao, Q., Wang, B., Duan, P., Zhang, B., Hong, Z.: Privacy-preserving multi-class support vector machine model on medical diagnosis. IEEE J. Biomed. Health Inform. **26**(7), 3342–3353 (2022)
7. Chilukoti, V.S.T.S.V., Hsu, S., Hei, X.: Privacy-preserving deep learning model for COVID-19 disease detection. arXiv preprint arXiv:2209.04445 (2022)
8. Dayan, I., et al.: Federated learning for predicting clinical outcomes in patients with COVID-19. Nat. Med. **27**(10), 1735–1743 (2021)
9. Deist, T.M., et al.: Distributed learning on 20 000+ lung cancer patients-the personal health train. Radiother. Oncol. **144**, 189–200 (2020)
10. Dosovitskiy, A., et al.: An image is worth 16×16 words: transformers for image recognition at scale. arXiv preprint arXiv:2010.11929 (2020)
11. Dou, Q., et al.: Federated deep learning for detecting COVID-19 lung abnormalities in CT: a privacy-preserving multinational validation study. NPJ Digit. Med. **4**(1), 60 (2021)
12. Dwork, C., Roth, A., et al.: The algorithmic foundations of differential privacy. Found. Trends® Theor. Comput. Sci. **9**(3–4), 211–407 (2014)
13. Festag, S., Spreckelsen, C.: Privacy-preserving deep learning for the detection of protected health information in real-world data: comparative evaluation. JMIR Format. Res. **4**(5), e14064 (2020)

14. Field, M., et al.: Infrastructure platform for privacy-preserving distributed machine learning development of computer-assisted theragnostics in cancer. J. Biomed. Inform. **134**, 104181 (2022)
15. Goldreich, O.: Secure multi-party computation. Manuscript. Preliminary version 78(110) (1998)
16. Gopalakrishnan, A., Kulkarni, N.P., Raghavendra, C., Manjappa, R., Honnavalli, P.B., Eswaran, S.: PriMed: private federated training and encrypted inference on medical images in healthcare. Available at SSRN 4196696 (2021)
17. Gostin, L.O., Levit, L.A., Nass, S.J., et al.: Beyond the HIPAA privacy rule: enhancing privacy, improving health through research (2009)
18. Hong, C., et al.: Privacy-preserving collaborative machine learning on genomic data using TensorFlow. In: Proceedings of the ACM Turing Celebration Conference-China, pp. 39–44 (2020)
19. Huang, Q.X., Yap, W.L., Chiu, M.Y., Sun, H.M.: Privacy-preserving deep learning with learnable image encryption on medical images. IEEE Access **10**, 66345–66355 (2022)
20. Islam, H., Alaboud, K., Paul, T., Rana, M.K.Z., Mosa, A.: A privacy-preserved transfer learning concept to predict diabetic kidney disease at out-of-network siloed sites using an in-network federated model on real-world data. In: AMIA Annual Symposium Proceedings, vol. 2022, p. 264. American Medical Informatics Association (2022)
21. Islam, T.U., Ghasemi, R., Mohammed, N.: Privacy-preserving federated learning model for healthcare data. In: 2022 IEEE 12th Annual Computing and Communication Workshop and Conference (CCWC), pp. 0281–0287. IEEE (2022)
22. Jarin, I., Eshete, B.: PRICURE: privacy-preserving collaborative inference in a multi-party setting. In: Proceedings of the 2021 ACM Workshop on Security and Privacy Analytics, pp. 25–35 (2021)
23. Javaid, M., Haleem, A., Singh, R.P., Suman, R., Rab, S.: Significance of machine learning in healthcare: features, pillars and applications. Int. J. Intell. Netw. **3**, 58–73 (2022)
24. Joshi, M., Pal, A., Sankarasubbu, M.: Federated learning for healthcare domain-pipeline, applications and challenges. ACM Trans. Comput. Healthc. **3**(4), 1–36 (2022)
25. Jumper, J., et al.: Highly accurate protein structure prediction with alphafold. Nature **596**(7873), 583–589 (2021)
26. Kaissis, G., Ziller, A., et al.: End-to-end privacy preserving deep learning on multi-institutional medical imaging. Nat. Mach. Intell. **3**(6), 473–484 (2021)
27. Kasyap, H., Tripathy, S.: Privacy-preserving decentralized learning framework for healthcare system. ACM Trans. Multimed. Comput. Commun. Appl. (TOMM) **17**(2s), 1–24 (2021)
28. Kaul, D., Raju, H., Tripathy, B.: Deep learning in healthcare. Deep Learning in Data Analytics: Recent Techniques, Practices and Applications, pp. 97–115 (2022)
29. Kerkouche, R., Acs, G., Castelluccia, C., Genevès, P.: Privacy-preserving and bandwidth-efficient federated learning: an application to in-hospital mortality prediction. In: Proceedings of the Conference on Health, Inference, and Learning, pp. 25–35 (2021)
30. Kreuzberger, D., Kühl, N., Hirschl, S.: Machine learning operations (MLOps): overview, definition, and architecture. arXiv preprint arXiv:2205.02302 (2022)
31. Lee, G.H., Shin, S.Y.: Federated learning on clinical benchmark data: performance assessment. J. Med. Internet Res. **22**(10), e20891 (2020)

32. Lins, S., Pandl, K.D., Teigeler, H., Thiebes, S., Bayer, C., Sunyaev, A.: Artificial intelligence as a service: classification and research directions. Bus. Inf. Syst. Eng. **63**, 441–456 (2021)
33. Liu, B., Ding, M., Shaham, S., Rahayu, W., Farokhi, F., Lin, Z.: When machine learning meets privacy: a survey and outlook. ACM Comput. Surv. (CSUR) **54**(2), 1–36 (2021)
34. Liu, P., Xu, X., Wang, W.: Threats, attacks and defenses to federated learning: issues, taxonomy and perspectives. Cybersecurity **5**(1), 1–19 (2022)
35. Loftus, T.J., et al.: Federated learning for preserving data privacy in collaborative healthcare research. Digit. Health **8**, 20552076221134456 (2022)
36. Ma, Z., et al.: Lightweight privacy-preserving medical diagnosis in edge computing. IEEE Trans. Serv. Comput. **15**(3), 1606–1618 (2020)
37. McMahan, B., Moore, E., Ramage, D., Hampson, S., y Arcas, B.A.: Communication-efficient learning of deep networks from decentralized data. In: Artificial Intelligence and Statistics, pp. 1273–1282. PMLR (2017)
38. Miotto, R., Wang, F., Wang, S., Jiang, X., Dudley, J.T.: Deep learning for healthcare: review, opportunities and challenges. Brief. Bioinform. **19**(6), 1236–1246 (2018)
39. Montenegro, H., Silva, W., Cardoso, J.S.: Privacy-preserving generative adversarial network for case-based explainability in medical image analysis. IEEE Access **9**, 148037–148047 (2021)
40. Nguyen, D.C., et al.: Federated learning for smart healthcare: a survey. ACM Comput. Surv. (CSUR) **55**(3), 1–37 (2022)
41. OECD: Robustness, security and safety (principle 1.4) (2023). https://oecd.ai/en/dashboards/ai-principles/P8
42. OECD: Transparency and explainability (principle 1.3) (2023). https://oecd.ai/en/dashboards/ai-principles/P7
43. Onesimu, J.A., Karthikeyan, J.: An efficient privacy-preserving deep learning scheme for medical image analysis. J. Inf. Technol. Manage. **12**(Special Issue: The Importance of Human Computer Interaction: Challenges, Methods and Applications), 50–67 (2020)
44. Ouyang, D., et al.: Video-based AI for beat-to-beat assessment of cardiac function. Nature **580**(7802), 252–256 (2020)
45. Park, S., Kim, G., Kim, J., Kim, B., Ye, J.C.: Federated split vision transformer for COVID-19 CXR diagnosis using task-agnostic training. arXiv preprint arXiv:2111.01338 (2021)
46. Paul, J., Annamalai, M.S.M.S., Ming, W., Al Badawi, A., Veeravalli, B., Aung, K.M.M.: Privacy-preserving collective learning with homomorphic encryption. IEEE Access **9**, 132084–132096 (2021)
47. Popescu, A.B., et al.: Privacy preserving classification of EEG data using machine learning and homomorphic encryption. Appl. Sci. **11**(16), 7360 (2021)
48. Qayyum, A., Qadir, J., Bilal, M., Al-Fuqaha, A.: Secure and robust machine learning for healthcare: a survey. IEEE Rev. Biomed. Eng. **14**, 156–180 (2020)
49. Ramachandram, D., Taylor, G.W.: Deep multimodal learning: a survey on recent advances and trends. IEEE Signal Process. Mag. **34**(6), 96–108 (2017)
50. Ravì, D., et al.: Deep learning for health informatics. IEEE J. Biomed. Health Inform. **21**(1), 4–21 (2016)
51. Sadilek, A., et al.: Privacy-first health research with federated learning. NPJ Digit. Med. **4**(1), 132 (2021)

52. Sarkar, E., Chielle, E., Gursoy, G., Chen, L., Gerstein, M., Maniatakos, M.: Scalable privacy-preserving cancer type prediction with homomorphic encryption. arXiv preprint arXiv:2204.05496 (2022)
53. Shehab, M., et al.: Machine learning in medical applications: a review of state-of-the-art methods. Comput. Biol. Med. **145**, 105458 (2022)
54. Singh, S., Shukla, K.: Privacy-preserving machine learning for medical image classification. arXiv preprint arXiv:2108.12816 (2021)
55. Soenksen, L.R., et al.: Integrated multimodal artificial intelligence framework for healthcare applications. NPJ Digit. Med. **5**(1), 149 (2022)
56. Srivastava, S.K., Singh, S.K., Suri, J.S.: Effect of incremental feature enrichment on healthcare text classification system: a machine learning paradigm. Comput. Methods Program. Biomed. **172**, 35–51 (2019)
57. Suriyakumar, V.M., Papernot, N., Goldenberg, A., Ghassemi, M.: Chasing your long tails: differentially private prediction in health care settings. In: Proceedings of the 2021 ACM Conference on Fairness, Accountability, and Transparency, pp. 723–734 (2021)
58. T'Jonck, K., Kancharla, C.R., Pang, B., Hallez, H., Boydens, J.: Privacy preserving classification via machine learning model inference on homomorphic encrypted medical data. In: 2022 XXXI International Scientific Conference Electronics (ET), pp. 1–6. IEEE (2022)
59. Vaswani, A., et al.: Attention is all you need. Adv. Neural Inf. Process. Syst. **30** (2017)
60. Vizitiu, A., Niţă, C.I., Puiu, A., Suciu, C., Itu, L.M.: Towards privacy-preserving deep learning based medical imaging applications. In: 2019 IEEE International Symposium on Medical Measurements and Applications (MeMeA), pp. 1–6. IEEE (2020)
61. Voigt, P., Von dem Bussche, A.: The EU General Data Protection Regulation (GDPR). A Practical Guide, 1st edn., vol. 10, no. 3152676, pp. 10–5555. Springer, Cham (2017). https://doi.org/10.1007/978-3-319-57959-7
62. Wang, Q., Zhou, Y.: Fedspl: federated self-paced learning for privacy-preserving disease diagnosis. Brief. Bioinform. **23**(1), bbab498 (2022)
63. Warnat-Herresthal, S., et al.: Swarm learning as a privacy-preserving machine learning approach for disease classification. BioRxiv, pp. 2020–06 (2020)
64. WHO: Who issues first global report on artificial intelligence (AI) in health and six guiding principles for its design and use (2021). https://www.who.int/news/item/28-06-2021-who-issues-first-global-report-on-ai-in-health-and-six-guiding-principles-for-its-design-and-use
65. Wibawa, F., Catak, F.O., Kuzlu, M., Sarp, S., Cali, U.: Homomorphic encryption and federated learning based privacy-preserving CNN training: COVID-19 detection use-case. In: Proceedings of the 2022 European Interdisciplinary Cybersecurity Conference, pp. 85–90 (2022)
66. Wiesenfeld, B.M., Aphinyanaphongs, Y., Nov, O.: AI model transferability in healthcare: a sociotechnical perspective. Nat. Mach. Intell. **4**(10), 807–809 (2022)
67. Wolff, J., et al.: Federated machine learning for a facilitated implementation of artificial intelligence in healthcare-a proof of concept study for the prediction of coronary artery calcification scores. J. Integr. Bioinform. **19**(4) (2022)
68. Xu, J., Glicksberg, B.S., Su, C., Walker, P., Bian, J., Wang, F.: Federated learning for healthcare informatics. J. Healthc. Inform. Res. **5**, 1–19 (2021)
69. Yan, R., et al.: Label-efficient self-supervised federated learning for tackling data heterogeneity in medical imaging. IEEE Trans. Med. Imaging (2023)

70. Yang, J., et al.: MedMNIST v2-a large-scale lightweight benchmark for 2D and 3D biomedical image classification. Sci. Data **10**(1), 41 (2023)
71. Yang, S., et al.: Towards the practical utility of federated learning in the medical domain. arXiv preprint arXiv:2207.03075 (2022)
72. Yue, Z., et al.: Privacy-preserving time-series medical images analysis using a hybrid deep learning framework. ACM Trans. Internet Technol. (TOIT) **21**(3), 1–21 (2021)
73. Zalonis, J., Armknecht, F., Grohmann, B., Koch, M.: Report: state of the art solutions for privacy preserving machine learning in the medical context. arXiv preprint arXiv:2201.11406 (2022)
74. Zerka, F., et al.: Blockchain for privacy preserving and trustworthy distributed machine learning in multicentric medical imaging (C-DistriM). IEEE Access **8**, 183939–183951 (2020)
75. Zhang, L., Xu, J., Vijayakumar, P., Sharma, P.K., Ghosh, U.: Homomorphic encryption-based privacy-preserving federated learning in IoT-enabled healthcare system. IEEE Trans. Netw. Sci. Eng. (2022)
76. Zhang, W., Tople, S., Ohrimenko, O.: Leakage of dataset properties in multi-party machine learning. In: USENIX Security Symposium, pp. 2687–2704 (2021)
77. Zhang, X., Ding, J., Wu, M., Wong, S.T., Van Nguyen, H., Pan, M.: Adaptive privacy preserving deep learning algorithms for medical data. In: Proceedings of the IEEE/CVF Winter Conference on Applications of Computer Vision, pp. 1169–1178 (2021)
78. Ziller, A., et al.: Privacy-preserving medical image analysis. arXiv preprint arXiv:2012.06354 (2020)

Self-supervised Predictive Coding with Multimodal Fusion for Patient Deterioration Prediction in Fine-Grained Time Resolution

Kwanhyung Lee[1], John Won[1,2], Heejung Hyun[1], Sangchul Hahn[1],
Edward Choi[3], and Joohyung Lee[1(✉)]

[1] AITRICS, Seoul, South Korea
{kwanlee9209,johnwon,alex,steve,chris}@aitrics.com
[2] POSTECH, Pohang-si, South Korea
johnwon@postech.ac.kr
[3] KAIST, Daejeon, South Korea
edwardchoi@kaist.ac.kr

Abstract. Accurate time prediction of patients' critical events is crucial in urgent scenarios where timely decision-making is important. Though many studies have proposed automatic prediction methods using Electronic Health Records (EHR), their coarse-grained time resolutions limit their practical usage in urgent environments such as the emergency department (ED) and intensive care unit (ICU). Therefore, in this study, we propose an hourly prediction method based on self-supervised predictive coding and multi-modal fusion for two critical tasks: mortality and vasopressor need prediction. Through extensive experiments, we prove significant performance gains from both multi-modal fusion and self-supervised predictive regularization, most notably in far-future prediction, which becomes especially important in practice. Our uni-modal/bi-modal/bi-modal self-supervision scored 0.846/0.877/0.897 (0.824/0.855/0.886) and 0.817/0.820/0.858 (0.807/0.81/0.855) with mortality (far-future mortality) and with vasopressor need (far-future vasopressor need) prediction data in AUROC, respectively.

Keywords: Self-supervision · predictive coding · multi-modal fusion · EHR · time-series prediction

1 Introduction

In the emergency department (ED) and intensive care unit (ICU), accurate time prediction of clinically critical events is crucial to make timely interventions for acutely deteriorating patients. Moreover, the early prediction of critical events enables precise prioritization and preparation for high-risk patients by an efficient resource allocation

---

K. Lee and J. Lee—Equal contribution.
J. Won—Work done while J. Won was an intern at AITRICS Inc.

© The Author(s), under exclusive license to Springer Nature Switzerland AG 2023
H. Chen and L. Luo (Eds.): TML4H 2023, LNCS 13932, pp. 41–50, 2023.
https://doi.org/10.1007/978-3-031-39539-0_4

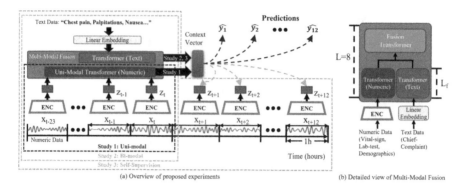

(a) Overview of proposed experiments   (b) Detailed view of Multi-Modal Fusion

Fig. 1. (a) Overview of proposed 1) uni-modal, 2) bi-modal, 3) self-supervised methods. All methods include predictions (supervised learning). t refers to the time when the prediction is made. (b) Fusion structure with L_f indicating where the fusion starts.

[16,18]. As a result, many studies have reported their early prediction systems using Electronic Health Records (EHR) [12,16,18].

Reported studies, however, make predictions in coarse-grained time resolution: 1) predicting vasopressor need within 24/48 h [2,17] and within 6–10 h [13] and 2) predicting mortality within 24/48 h [17] and 3) within the whole hospitalization period [3,16]. However, prediction over a coarse-grained time resolution can be impractical where timely decision-making and rapid intervention are crucial. In this study, therefore, we aim to 1) make an hourly prediction over the future 12 h and 2) enhance the far-future (early) prediction by adding static features to the EHR time-series data [18].

Several studies suggest utilizing multi-modal EHR data in coarse-grained time resolution, such as Wang *et al.* [16], which predicts future mortality with physiological index, treatment records, and hospitalization records, or Suresh *et al.* [13], which predicts intervention needs with demographic data, vital signs/lab tests, and clinical notes. However, the benefit of additional modalities was uncertain, since these studies report the performance of the multi-modal model without comparing it to the performance of the individual (uni-modal) model.

Our main contributions are as follows; 1) we propose a novel fine-grained time deterioration prediction method for two representative critical events in ED and ICU, i.e., mortality and vasopressor need; 2) we show our future encoding method with additional normalization is important in our self-supervised predictive regularization for fine-grained deterioration prediction (Fig. 2); 3) through an extensive experiment, we show that both multi-modal fusion and our self-supervised predictive regularization improves the predictive performance, especially the far-future (early) prediction, which is crucial but more challenging than near-future prediction [4,18].

2 Methods

Figure 1 illustrates the overall scheme of our network. In this paper, we propose a novel deterioration prediction model for a fine-grained time resolution. To this end, we grad-

Table 1. Data statistics with patient numbers for mortality prediction and vasopressor need prediction tasks.

Tasks	Mortality	Vasopressor
Data Split	Train Test	Train Test
Positive Subjects	2544/262	5827/606
Negative Subjects	24492/2836	21941/2580

ually add features with the best performance in the following order; 1) we select a uni-modal model to learn EHR numeric data; 2) we compare bi-modal (numeric + text data) fusion strategies; 3) we compare various joint self-supervised learning (SSL) strategies; and 4) we compare various encoding methods for the future EHR numeric data to perform joint SSL. For each study, we select the best-performing one and fix it for the remaining studies to assess the efficacy of the added component, e.g., additional modality or SSL loss (Fig. 3). We conduct every study for both mortality and vasopressor need prediction data. At the end of the last study, therefore, we propose our deterioration prediction model for a fine-grained time resolution with the accumulated components. For a fair comparison, we fix all Transformer-based models to equally have 8 transformer layers, 4 multi-heads, and 256 feature dimensions (Fig. 1(b)) (Table 1).

2.1   Electronic Health Record Data

To simulate an urgent hospital environment, we used the MIMIC-ED and MIMIC-IV (Medical Information Mart for Intensive Care in Emergency Department and IV) datasets [7,8]. Since both datasets share the same patients, we merged chief complaints from MIMIC-ED (text data) and a total of 18 different time-series features from MIMIC-IV (numeric data), i.e., vital signs, lab-test results, and demographic features (age and gender). Vital-sign includes heart rate, respiration rate, and 4 other items. Lab-test result include Hematocrit, Platelet, and 8 more items [12]. More detailed information about the selected features and their importance for our prediction tasks are summarized in Appendix section A.1 and A.2. We labeled the occurrence of mortality and vasopressor usage in binary. The sampling frequency for the time-series data is 1 h, and we applied carry-forward imputation (most recent value of the past) for missing features. For vital-sign and lab-test features, we applied min-max normalization using the minimum and maximum values from the entire training dataset. The input time length varies from 3 to 24 h to 1) challenge the prediction for patients shortly after admission and 2) simulate varying ED environments [6]. We used zero-padding to fix the input data length as 24 h and only considered patients who had ICU stays of 15 to 1440 h.

To select the best-performing model for the four studies (Table 2), we used the averaged validation area under the receiver operating characteristic (AUROC) from 5-fold cross-validation (CV). To assess the efficacy of the additional component, i.e. text data and SSL, we compared the averaged test AUROC from the 5-fold CV of the best-performing uni-modal, bi-modal, and the model trained with SSL (Fig. 3).

2.2   Uni-Modal Model for EHR Numeric Data

We explored four different models: GRU-D [1], LSTM, Transformer [15], and Graph Transformer [3,9]. All four models map time-series numeric data (vital-signs, lab tests, demographics) $x_{\leq t} \in \mathbb{R}^{18 \times T_i}$, ($T_i = 24$ in our study) into a context vector $c_t \in \mathbb{R}^{256}$, which is then mapped to 12 probabilities for our 12-h fine-grained time prediction. For mapping, we use 12 distinct 2-layer Multilayer perceptron (MLP) with batch normalization and ReLU non-linearities between the 2 linear layers, followed by a sigmoid function. Both GRU-D and LSTM receive the raw input $x_{\leq t}$, whereas both Transformer and Graph Transformer receive the encoded input $z_{\leq t}$, which is $x_{\leq t}$ encoded by the 2-layer MLP with Layer Normalization (LN) and ReLU activation, to output the context vector c_t (CLS token vector) (Fig. 1(a)).

2.3   Bi-modal Fusion Strategy for EHR Text Data

Alike the best-performing uni-modal model (Table 2), we use the vanilla Transformer with BERT tokenization [5] for EHR text data. We fuse the outcomes of the L_f-th layer of the text and numeric Transformers (Fig. 1(b)); we refer the *early* and *mid* fusion to the fusions that occur after the 0-th (before Transformer) and 4-th layer of the Transformers of text and numeric data (Fig. 1(b)). Note that we rigorously explore the *early* and *mid* fusion due to the poor performance of the *late* fusion (fusion after 9-th layer) during our preliminary experiment. Moreover, we experimented with three different types of fusion methods: 1) Multimodal Bottleneck Transformer (MBT) [10], 2) Multimodal-Transformer (MT) [10], 3) Bi-Cross Modal Attention Transformer (BCMAT) [14]. MBT, MT, and BCMAT respectively utilize the fusion bottleneck (FSN) tokens [10], concatenation, and attention fusion after the L_f-th layer. Specifically, MBT creates and lets two Transformers share four new FSN tokens after the L_f-th layer of the text and numeric Transformers. MT concatenates the outcomes of the L_f-th layer of both Transformers. BCMAT uses two parallel attention fusions; after L_f-th layer, one Transformer uses its outcome as both the key and value and the outcome of the other Transformer as the query, and vice versa for the other fusion.

2.4   Self-supervised Regularization

For time-series data, Oord *et al.* [11] introduced Contrastive Predictive Coding (CPC) which self-supervises networks to encourage capturing global information, i.e. 'slow

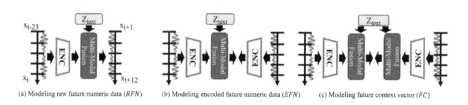

(a) Modeling raw future numeric data (*RFN*)   (b) Modeling encoded future numeric data (*EFN*)   (c) Modeling future context vector (*FC*)

Fig. 2. Three different methods to encode the future numeric data (x_{t+k}) for self-supervision. (b) is the default method used in Sect. 2.4 and illustrated in Fig. 1a.

feature', using Noise Contrastive Estimation (NCE). Moreover, Wanyan *et al.* [17] and Zang *et al.* [19] proposed to regularize networks by adding a SSL loss to the supervision loss. Motivated by these studies, we regularize our bi-modal network using CPC whose loss can be described as Eq. 1:

$$\mathscr{L}_{NCE} = -log\frac{exp(z_{t+k}^T W_k c_t)}{\sum_j exp(z_j^T W_k c_t)} \tag{1}$$

$$\mathscr{L}_{cosine} = -\frac{z_{t+k}^T W_k c_t}{\|z_{t+k}\|_2 \cdot \|W_k c_t\|_2} \tag{2}$$

$$\mathscr{L}_{l_2} = \|z_{t+k} - W_k c_t\|_2 \tag{3}$$

However, since we add supervision loss to SSL loss (NCE), we assume the model will not converge to a trivial solution (model collapse) even if the SSL loss does not contain negative pairs. Therefore, we also implement cosine (Eq. 2) and L_2 loss (Eq. 3) alongside NCE. In the equations above, c_t refers to the context vector (Fig. 1(a)) from the bi-modal fusion Transformer. We use linear transformation $W_k c_t$ for SSL prediction where different W_k is used for different time step k. In this study, we use 12 distinct W_k to concurrently predict 12 encoded future numeric z_{t+1}, z_{t+2}, ..., z_{t+12}. Note that, the encoder for the past numeric $x_{\leq t}$ encodes 12 distinct future numeric $x_{>t}$ to $z_{>t}$ as well. Since we maximize the mutual information (MI) between linearly transformed context vectors and 12 distinct $z_{>t}$, which share the same encoder, the encoder is encouraged to learn the information shared across all time points; we assume this 'slow feature' encourages far-future prediction. For L_2 loss of MT_{early} (mortality prediction), we explore additional LN to normalize c_t, because L_2 loss from unnormalized c_t is large in its value compared to supervision loss (MBT_{early} does not need additional LN since its context vector is already normalized).

2.5 Encoding Future Numeric Data for Self-supervision

The original CPC paper proposes to maximize MI between context vector c_t and encoded future numeric z_{t+k} (Fig. 2(b)) instead of using raw future numeric x_{t+k} (Fig. 2(a)). The aim of the original CPC paper is to avoid modeling the high dimensional distribution of the raw data x_{t+k}. Since our raw data x_{t+k} has lower dimensions than the encoded data z_{t+k}. Therefore, we hypothesized that modeling the raw future numeric may outperform modeling the encoded future numeric (Fig. 2(a)). Lastly, we also experimented Fig. 2(c) which encourages similarity between the context vectors of the past and the future assuming that using the time-aggregated context vector for SSL may outperform the others. We compare the performance of these three SSL structures (Fig. 2) in Sect. 3.2.

3 Results and Discussion

In this section, we analyze the performances of different 1) models to learn EHR numeric data, 2) strategies to fuse the encoded EHR text data and numeric data, 3) self-supervision loss, and 4) how to use the future EHR numeric data for self-supervision.

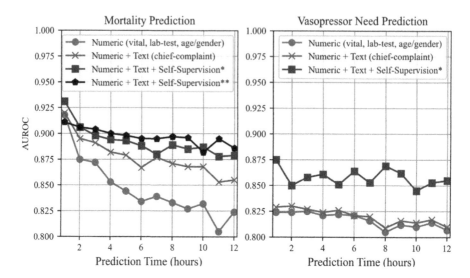

Fig. 3. Average test AUROC of mortality (left) and vasopressor need (right) prediction using the best-performing model from each study. * and ** indicates the best performing model from Sect. 2.4 and Sect. 2.5, respectively. Average AUROC are 0.8463, 0.877, 0.8925, 0.8971 for mortality task and 0.8166, 0.8203, 0.858 for vasopressor task (from top to bottom in the legend box).

We conduct these four studies to predict mortality and vasopressor need independently and illustrate the result in Table 2.

3.1 Transformer Works Better Than Other Alternatives for Learning EHR Numeric Data

As shown in Table 2, the vanilla Transformer outperforms all other alternatives to predict mortality and vasopressor need in a fine-grained time course. Note that the vanilla Transformer excels over the Graph Transformer suggesting that learning temporal relationships is more important than learning inter-feature relationships. All four models show gradual degradation in performance when predicting further in the future, which reflects the difficulty in far-future prediction.

For fusing EHR text data to EHR numeric, early fusion outperforms other strategies. Particularly, feature concatenation (MT) benefits mortality prediction the most, whereas MBT improves the vasopressor need task the most.

3.2 Self-supervised Predictive Regularization Using L_2 Loss with Normalized Context Vector Is Crucial

As shown in Table 2, SSL regularization using $\mathscr{L}_2$ loss with normalized context vector c_t by LN performs the best for both prediction tasks. Note that the performance gap between $MT_{early}L_{2EFN}$ and $MT_{early*}L_{2EFN}$ indicates the importance of context vector

Table 2. Validation AUROC of 1) uni-modal, 2) bi-modal, 3) self-supervision loss, and 4) different future numeric encoding methods for self-supervision. The range a–$a+1$ indicates future prediction's hourly range. Best performing option in average AUROC (bold) is selected and applied to the remaining studies to assess the efficacy of an additional feature. $*$ indicates additional normalization (Sect. 2.4). RFN, EFN, and FC refer to the different future numeric data encoding methods for SSL (Fig. 2).

Models	0–1	1–2	2–3	3–4	4–5	5–6	6–7	7–8	8–9	9–10	10–11	11–12	Avg.
Mortality Prediction (AUROC)													
GRU-D	0.897	0.867	0.852	0.838	0.831	0.822	0.812	0.798	0.788	0.801	0.786	0.78	0.823
LSTM	0.918	0.886	0.868	0.86	0.852	0.84	0.841	0.834	0.827	0.819	0.817	0.809	0.848
Transformer	**0.921**	**0.892**	**0.877**	**0.865**	0.856	0.846	0.84	0.837	0.827	**0.823**	**0.823**	**0.821**	**0.852**
Graph Transformer	0.903	0.885	0.875	**0.865**	**0.858**	**0.847**	**0.842**	**0.839**	**0.832**	0.822	**0.823**	0.819	0.851
MBT_{early}	0.91	0.887	0.881	0.876	0.872	**0.865**	0.859	0.86	**0.858**	0.854	0.851	**0.857**	0.869
MBT_{mid}	0.911	0.888	0.882	0.875	**0.873**	0.863	0.858	**0.861**	0.854	**0.858**	**0.857**	**0.857**	0.87
MT_{early}	0.913	0.897	0.888	**0.88**	**0.873**	**0.865**	**0.863**	0.86	0.854	0.856	0.852	0.848	**0.871**
MT_{mid}	**0.918**	0.896	0.885	0.876	0.871	0.862	0.858	0.854	0.846	0.846	0.843	0.839	0.866
$BCMAT_{early}$	0.916	**0.9**	**0.89**	0.873	0.868	0.863	0.853	0.855	0.844	0.84	0.84	0.838	0.865
$BCMAT_{mid}$	0.903	0.888	0.876	0.869	0.864	0.854	0.847	0.848	0.841	0.842	0.845	0.842	0.86
$MT_{early}NCE_{EFN}$	0.9	0.885	0.875	0.867	0.868	0.853	0.848	0.851	0.844	0.837	0.836	0.822	0.857
$MT_{early}Cosine_{EFN}$	0.91	0.902	0.89	0.872	0.884	0.876	0.84	0.86	0.865	0.875	0.855	0.839	0.872
$MT_{early}L2_{EFN}$	0.696	0.562	0.75	0.709	0.687	0.722	0.736	0.574	0.647	0.667	0.676	0.696	0.677
$MT_{early*}L2_{EFN}$	**0.926**	**0.907**	**0.898**	**0.885**	**0.89**	**0.883**	**0.871**	**0.879**	**0.877**	**0.876**	**0.88**	0.871	**0.887**
$MT_{early*}L2_{RFN}$	**0.902**	**0.904**	**0.895**	**0.887**	**0.895**	**0.892**	**0.886**	**0.885**	**0.883**	**0.875**	**0.891**	**0.878**	**0.889**
$MT_{early*}L2_{FC}$	0.921	0.894	0.885	0.874	0.868	0.858	0.854	0.849	0.847	0.841	0.836	0.829	0.863
Vasopressor Need Prediction (AUROC)													
GRU-D	0.819	0.815	0.813	**0.813**	**0.811**	0.805	0.801	0.8	0.797	0.794	**0.795**	0.789	0.804
LSTM	0.814	0.81	0.808	0.806	0.802	0.8	0.795	0.794	0.791	0.79	0.785	0.782	0.798
Transformer	**0.818**	**0.817**	**0.815**	**0.813**	0.81	**0.808**	**0.802**	**0.802**	**0.798**	**0.797**	0.793	**0.791**	**0.805**
Graph Transformer	0.808	0.808	0.804	0.803	0.8	0.799	0.793	0.79	0.789	0.787	0.784	0.782	0.796
MBT_{early}	**0.826**	**0.824**	**0.82**	**0.817**	**0.815**	**0.811**	**0.806**	**0.805**	**0.802**	**0.799**	0.797	**0.794**	**0.81**
MBT_{mid}	0.821	0.819	0.816	0.815	0.812	0.809	0.805	0.804	0.799	0.798	0.794	0.792	0.807
MT_{early}	0.814	0.817	0.815	0.814	0.809	0.808	0.804	0.802	0.798	0.798	0.795	0.793	0.806
MT_{mid}	0.819	0.82	0.818	0.816	0.813	0.81	0.805	0.804	0.799	0.796	0.793	0.791	0.807
$BCMAT_{early}$	0.811	0.813	0.815	0.806	0.814	0.81	0.796	0.808	0.788	0.796	**0.802**	0.791	0.804
$BCMAT_{mid}$	0.819	0.817	0.814	0.811	0.809	0.807	0.802	0.801	0.797	0.796	0.793	0.791	0.805
$MBT_{early}NCE_{EFN}$	0.807	0.806	0.808	0.804	0.807	0.803	0.803	0.798	0.793	0.797	0.792	0.797	0.8
$MBT_{early}Cosine_{EFN}$	0.812	0.808	0.811	0.808	0.811	0.809	0.807	0.801	0.798	0.802	0.796	0.793	0.805
$MBT_{early}L2_{EFN}$	**0.871**	**0.845**	**0.851**	**0.856**	**0.841**	**0.857**	**0.844**	**0.867**	**0.852**	**0.836**	**0.842**	**0.843**	**0.851**
$MBT_{early}L2_{RFN}$	0.834	0.827	0.834	0.828	0.827	0.833	0.83	0.838	0.838	0.823	0.819	0.827	0.829
$MBT_{early}L2_{FC}$	0.819	0.815	0.816	0.814	0.81	0.808	0.802	0.802	0.797	0.796	0.793	0.791	0.805

normalization. Note that the L2 loss yields comparatively larger values than the other auxiliary losses, i.e., cosine and NCE. As a result, incorporating an additional normalization process balance the auxiliary L2 loss with supervised loss, thus improving the model performance. Moreover, SSL with encoded future data (Fig. 2(b)), which is introduced in the original CPC paper does not always outperform other alternatives, which we partly connect with the low dimensionality of raw future numeric in Sect. 2.5.

3.3 Both Bi-modal Fusion and Self-supervised Predictive Regularization Improves Mortality and Vasopressor Need Prediction

After selecting the best option using validation AUROC for 1) a unimodal network, 2) bimodal fusion strategy, and 3) SSL method respectively, we used the test AUROC to compare their performances. As shown in Fig. 3, both EHR text supplementation and SSL regularization improve the predictive performance of both tasks, i.e., mortality prediction and vasopressor need prediction. Specifically, adding EHR text data to EHR numeric by bi-modal fusion improves overall/far-future (11–12h) prediction of the uni-modal model (baseline) by 0.031/0.031 in mortality prediction and by 0.004/0.003 in vasopressor need prediction. Additional SSL loss further improves overall/far-future prediction of the bi-modal model by 0.020/0.031 in mortality prediction and by 0.038/0.045 in vasopressor need prediction. Note that for the baseline unimodal method, mortality predictive performance degrades much as prediction time gets further in the future, unlike vasopressor need prediction. Though both bimodal fusion and SSL regularization improve the overall predictive performance for mortality prediction, they yield more improvement as prediction time gets further in the future. However, for vasopressor need prediction, the performance gap between near-future prediction and far-future prediction is small, and self-supervision helps overall prediction accuracy much more than text data supplementation.

4 Conclusion

This paper proposes a novel hourly deterioration prediction model for urgent patients in the ED/ICU. With extensive experiments, we show that both multi-modal fusion and self-supervised predictive regularization effectively improve the performance of mortality and vasopressor need prediction in a fine-grained time resolution; in mortality prediction, both multi-modal fusion and SSL regularization specifically improve the far-future prediction. For vasopressor need prediction, SSL improves not only the far-future prediction but also the overall prediction. In addition, we show the importance of context vector normalization for L_2 loss in SSL predictive coding regularization. We believe our method will advance timely intervention and effective resource allocation in the ED/ICU with the improved and thus more trustworthy prediction of patient's critical events.

A Appendix

A.1 Selected Features

Our selected six vital-sign data includes heart rate, respiration rate, blood pressure (diastolic and systolic), temperature, and pulse oximetry. The selected 10 lab-tests include Hematocrit, Platelet, WBC, Bilirubin, pH, HCO3, Creatinine, Lactate, Potassium, and Sodium [12].

A.2 Feature Importance

For more detailed information about each numeric feature's influence on the prediction decision, we further calculated the contribution of vital-signs and lab-test on mortality and vasopressor use prediction (Fig. 4).

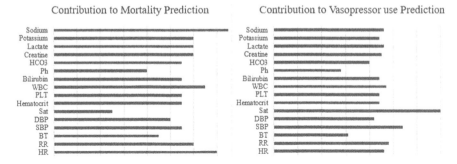

Fig. 4. Contribution of vital signal and lab-test features for mortality and vasopressor use prediction with Uni-modal Transformer. We used the Integrated Gradients and averaged 12 future predictions.

A.3 Model Complexity

(See Table 3)

Table 3. Number of parameters on each model. ∗ indicates any of the three different loss types of $NCE, Cosine$, or L_2

Models	Number of Parameters (M)
GRU-D	1.03
LSTM	0.41
Transformer	6.06
Graph Transformer	7.16
MBT_{early}	19.74
MBT_{mid}	19.74
MT_{early}	13.74
MT_{mid}	16.64
$BCMAT_{early}$	17.64
$BCMAT_{mid}$	18.68
$MT_{early*EFN}$	14.53
$MT_{early*RFN}$	14.53
$MT_{early*FC}$	13.74
$MBT_{early*EFN}$	21.31
$MBT_{early*RFN}$	21.31
$MBT_{early*FC}$	19.74

References

1. Che, Z., Purushotham, S., Cho, K., Sontag, D., Liu, Y.: Recurrent neural networks for multivariate time series with missing values. Sci. Rep. **8**(1), 1–12 (2018)
2. Choi, A., Chung, K., Chung, S.P., Lee, K., Hyun, H., Kim, J.H.: Advantage of vital sign monitoring using a wireless wearable device for predicting septic shock in febrile patients in the emergency department: a machine learning-based analysis. Sensors **22**(18), 7054 (2022)
3. Choi, E., et al.: Graph convolutional transformer: Learning the graphical structure of electronic health records. arXiv preprint arXiv:1906.04716 (2019)
4. Danilatou, V., et al.: Outcome prediction in critically-ill patients with venous thromboembolism and/or cancer using machine learning algorithms: external validation and comparison with scoring systems. Int. J. Mol. Sci. **23**(13), 7132 (2022)
5. Devlin, J., Chang, M.W., Lee, K., Toutanova, K.: BERT: bidirectional encoder representations from transformers (2016)
6. Henriksen, D.P., Brabrand, M., Lassen, A.T.: Prognosis and risk factors for deterioration in patients admitted to a medical emergency department. PLoS ONE **9**(4), e94649 (2014)
7. Johnson, A., Bulgarelli, L., Pollard, T., Celi, L.A., Mark, R., Horng, S.: Mimic-iv-ed
8. Johnson, A., Bulgarelli, L., Pollard, T., Horng, S., Celi, L.A., Mark, R.: Mimic-iv. PhysioNet (2020). https://physionet.org/content/mimiciv/1.0/. Accessed 23 Aug 2021
9. Lee, K., Jeong, H., Kim, S., Yang, D., Kang, H.C., Choi, E.: Real-time seizure detection using EEG: a comprehensive comparison of recent approaches under a realistic setting. arXiv preprint arXiv:2201.08780 (2022)
10. Nagrani, A., Yang, S., Arnab, A., Jansen, A., Schmid, C., Sun, C.: Attention bottlenecks for multimodal fusion. In: Advances in Neural Information Processing Systems, vol. 34, pp. 14200–14213 (2021)
11. van den Oord, A., Li, Y., Vinyals, O.: Representation learning with contrastive predictive coding. arXiv preprint arXiv:1807.03748 (2018)
12. Sung, M., et al.: Event prediction model considering time and input error using electronic medical records in the intensive care unit: retrospective study. JMIR Med. Inform. **9**(11), e26426 (2021)
13. Suresh, H., Hunt, N., Johnson, A., Celi, L.A., Szolovits, P., Ghassemi, M.: Clinical intervention prediction and understanding with deep neural networks. In: Machine Learning for Healthcare Conference, pp. 322–337. PMLR (2017)
14. Tsai, Y.H.H., Bai, S., Liang, P.P., Kolter, J.Z., Morency, L.P., Salakhutdinov, R.: Multimodal transformer for unaligned multimodal language sequences. In: Proceedings of the Conference. Association for Computational Linguistics. Meeting, vol. 2019, p. 6558. NIH Public Access (2019)
15. Vaswani, A., et al.: Attention is all you need. In: Advances in Neural Information Processing Systems, vol. 30 (2017)
16. Wang, Y., Lan, Y.: Multi-view learning based on non-redundant fusion for ICU patient mortality prediction. In: 2022 IEEE International Conference on Acoustics, Speech and Signal Processing (ICASSP), ICASSP 2022, pp. 1321–1325. IEEE (2022)
17. Wanyan, T., et al.: Contrastive learning improves critical event prediction in Covid-19 patients. Patterns **2**(12), 100389 (2021)
18. Wu, M., Ghassemi, M., Feng, M., Celi, L.A., Szolovits, P., Doshi-Velez, F.: Understanding vasopressor intervention and weaning: risk prediction in a public heterogeneous clinical time series database. J. Am. Med. Inform. Assoc. **24**(3), 488–495 (2017)
19. Zang, C., Wang, F.: Scehr: supervised contrastive learning for clinical risk prediction using electronic health records. In: 2021 IEEE International Conference on Data Mining (ICDM), pp. 857–866 (2021). https://doi.org/10.1109/ICDM51629.2021.00097

Safe Exploration in Dose Finding Clinical Trials with Heterogeneous Participants

Isabel Chien[1]([✉]), Javier Gonzalez Hernandez[2], and Richard E. Turner[1]

[1] University of Cambridge, Cambridge, UK
ic390@cam.ac.uk
[2] Microsoft Research, Cambridge, UK

Abstract. In drug development, early phase dose-finding clinical trials are carried out to identify an optimal dose to administer to patients in larger confirmatory clinical trials. Standard trial procedures do not optimize for participant benefit and do not consider participant heterogeneity, despite consequences to the health of participants and downstream impacts to under-represented population subgroups. Additionally, many newly investigated drugs do not obey modelling assumptions made in common dose-finding procedures. We present Safe Allocation for Exploration of Treatments (SAFE-T), a procedure for adaptive dose-finding that works well with small samples sizes and improves the utility for heterogeneous participants while adhering to safety constraints for treatment arm allocation. SAFE-T flexibly learns models for drug toxicity and efficacy without requiring strong prior assumptions and provides final recommendations for optimal dose by participant subgroup. We provide a preliminary evaluation of SAFE-T on a comprehensive set of realistic synthetic dose-finding scenarios, illustrating the improved performance of SAFE-T with respect to safety, utility, and dose recommendation accuracy across heterogeneous participants against a comparable baseline method.

1 Introduction

New drugs and treatments are generally first investigated in early phase dose-finding studies, which aim to assess safety and provide recommended dose levels for future study. Dose-finding trials often include a small number of participants due to safety concerns and difficulties with participant recruitment. Increasingly, researchers are promoting adaptive trial methods (in contrast to rule-based methods), where trial parameters may change based on ongoing participant outcomes, which can improve both the efficiency of a trial and the outcomes experienced by trial participants [19,27].

Dose-finding studies commonly use the rule-based 3+3 method, which has been criticized for its inefficiency [13], or the adaptive continual reassessment

---

Supplementary Information The online version contains supplementary material available at https://doi.org/10.1007/978-3-031-39539-0_5.

method (CRM), which requires a pre-selected parametric model and strict prior assumptions [30]. These methods assume that the patient population is homogeneous and do not account for possible variations in drug toxicity and efficacy due to patient heterogeneity. Due to these assumptions, alongside persisting inequalities in subject selection for clinical trials [5,21], optimal drug doses derived from such trials are often not generalizable to women, who remain under-represented in early-phase trials [15]. Multiple analyses have shown that women experience a far greater risk of adverse drug effects across all drug classes as compared to men [26,32]. Recent work has found that race and ethnicity can also impact drug response and has highlighted the inadequate understanding of drug safety and efficacy across under-studied racial and ethnic populations [6,18]. In addition to the issue of participant heterogeneity, newly developed drugs may not adhere to standard prior assumptions, particularly that dose-toxicity and dose-efficacy levels are monotonically increasing. Innovative therapies may follow plateauing or unimodal efficacy functions [28,31].

While complex factors surrounding inequality in clinical trials must be addressed to improve outcomes for under-represented populations, we make a small contribution with our adaptive dose-finding procedure, Safe Allocation for Exploration of Treatments (SAFE-T). SAFE-T models toxicity and efficacy dose-response curves with multi-output Gaussian processes that capture variations between participant subgroups and are flexible to different dose-response shapes. Using these dose-response estimates, SAFE-T balances safety and exploration in allocation of doses to trial participants, resulting in both improved participant utility and final dose recommendation accuracy.

Related Work. Recent works in machine learning for dose-finding trials have mostly concentrated on multi-armed bandit methods [2], with some including explicit safety constraints [14,20,29]. These works consider dose-finding scenarios with monotonically increasing toxicity and efficacy curves, with [2,20] also providing separate algorithms for plateauing efficacy curves. [14] propose methods that address heterogeneous populations with pre-defined subgroups. We note that [14] addresses a dose-finding setting most similar to ours; however, their method includes an un-safe burn-in period and uses a parametric model that is not flexible to different toxicity curve shapes. The problem of safe exploration has been addressed more generally in the context of bandits [11] and Bayesian optimization with Gaussian processes [22,23]. The issue of fair exploration (with respect to population subgroups) for bandit algorithms has also been investigated in [3,17].

2   Problem Statement

2.1   Dose-Finding Setup

We examine early phase dose-finding trials where N trial participants are sequentially, at timestep t, allocated one of K discrete doses, indexed by $k \in \mathbb{K} = \{1, ..., K\}$, with dose levels $d_k \in \mathbb{D}$ representing the dosage values. Participants belong to a known subgroup $s \in \mathbb{S} = \{1, ..., S\}$, arriving at rates π_s. For doses d, $g_s(d)$ defines true toxicity probabilities and $f_s(d)$ defines true efficacy probabilities by subgroup s. At each timestep, a participant of subgroup $h_t \in \mathbb{S}$ is allocated a dose of index $m_t \in \mathbb{K}$ based on a specified *selection rule*. Following dose allocation, we observe the binary toxicity outcome $Y_t \sim Ber(g_{h_t}(d_{m_t}))$, with $Y_t = 1$ indicating an adverse reaction; and the binary efficacy outcome $X_t \sim Ber(f_{h_t}(d_{m_t}))$, with $X_t = 1$ indicating effective treatment.

Trials using a model-based methodology will aim to accurately estimate the dose-toxicity response relationship, $\hat{g}_s(d)$, which describes the probability of a toxic event, and sometimes the dose-efficacy response relationship, $\hat{f}_s(d)$, which describes the probability of effective treatment, at doses $d \in \mathbb{D}$ for subgroups $s \in \mathbb{S}$. Adverse side effects are categorized as toxic events and patient responses are categorized as effective treatment based on previous clinical knowledge and will typically be formalized prior to implementation of a clinical trial. Trial practitioners specify a target toxicity threshold (TTL), τ_T, which is the highest acceptable probability of a toxic event. Similarly, τ_E is also specified as the lowest acceptable probability of efficacy for a dose. A dose d for a patient in subgroup s is thus in an acceptable safe range when $g_s(d) \leq \tau_T$ and $f_s(d) \geq \tau_E$.

At the conclusion of a trial, a *recommendation rule* is used to select a final *dose recommendation* $\hat{d}_{s,N}$, which should be equivalent to the *optimal dose* d_s^* for each subgroup s. These doses would be examined in larger, downstream clinical trials that focus on determining the efficacy of drugs. Commonly used dose-finding methods select the maximum tolerated dose as the optimal dose, defined as $\hat{d}_{s,N} = \arg\max_{d_k : \hat{g}_s(d) \leq \tau_T} \hat{g}_s(d)$. However, this recommendation rule would fail in cases where the dose-efficacy curve plateaus or is unimodal. SAFE-T provides an alternative method for selecting optimal doses that works well to determine optimal doses across differing curve shapes.

2.2 Problem Constraints

We discuss the realistic constraints and objectives of an algorithm that can be used for dose-finding trials. Dose-finding trials may incorporate **heterogeneous patients** belonging to differing pre-defined subgroups. Each subgroup may adhere to transformed dose-toxicity and dose-efficacy curves, although it may not be previously known what the difference may be [25]. As such, we desire an algorithm that can flexibly learn differences between in the dose-response relationships of patient subgroups. Our algorithm should provide **accurate final dose recommendations** across subgroups, where $\hat{d}_{s,N} = d_s^*, \forall s \in \mathbb{S}$. The **safety** of trial participants is of paramount concern. We aim to minimize safety constraint violations and also avoid using a burn-in period as done in some multi-armed bandit solutions for dose-finding [14,20], where the algorithm initializes by selecting each dose in succession regardless of safety. While standard dose-finding procedures do not consider **efficacy** for trial participants, we incorporate it into our methods in order to improve participant outcomes. Recent work in ML has argued that in healthcare settings, we should aim to **maximize expected utility** rather than institute fairness constraints or objectives that may reduce utility [16]. As such, we do not provide explicit fairness constraints or objectives in our methodology, but assess our performance with respect to utility and safety across participant subgroups. The majority of early phase trials include a **small sample size**, often below 50 participants [1,8]. An effective algorithm must be able to learn dose-response relationships for toxicity and efficacy efficiently, with few samples.

3   The SAFE-T Algorithm

Algorithm 1 SAFE-T Algorithm

1: **Input:** patient subgroups $\mathbb{S}$, dose indices $\mathbb{K}$, number of patients N, safe expansion timesteps N_0, toxicity threshold τ_T, efficacy threshold τ_E, GP prior for multi-output toxicity function $\hat{g}(d)$, GP prior for multi-output efficacy function $\hat{f}(d)$, utility function $U(p_e, p_t)$,

2: **Initialize:** $t \leftarrow 1$, $\mathbb{B}_{s,0} \leftarrow \emptyset \forall s \in \mathbb{S}$

3: **while** $t \leq N$ **do**

4:   $b = \begin{cases} 0, & \text{if } \mathbb{B}_{h_t,t-1} = \emptyset \\ \max(\mathbb{B}_{h_t,t-1}), & \text{otherwise} \end{cases}$

5:   $\mathbb{A}_{h_t,t} \leftarrow \mathbb{B}_{h_t,t-1} \bigcup \{b+1\}$

6:   **if** $t < N_0$ **then**

7:     $\mathbb{M}_{h_t,t} \leftarrow \{k \in \mathbb{A}_{h_t,t} \,|\, \mu_{\hat{g}_{h_t}(d_k)} \leq \tau_t\}$

8:     **if** $\max(\mathbb{M}_{h_t,t}) \notin \mathbb{B}_{h_t,t-1}$ **then**

9:       $m_t \leftarrow \max(\mathbb{M}_{h_t,t})$

10:    **else**

11:      $m_t \leftarrow \arg\max_{k \in \mathbb{M}_{h_t,t}} c_t(d_k)$

12:    **end if**

13:   **else**

14:     $\mathbb{M}_{h_t,t} \leftarrow \{k \in \mathbb{A}_{h_t,t} \,|\, u_t(d_k) \leq \tau_t\}$

15:     $m_t \leftarrow \arg\max_{d_k \in \mathbb{M}_{h_t,t}} EI(d_k)$

16:   **end if**

17:   Observe outcomes X_t, Y_t

18:   Update $\hat{g}(d)$, $\hat{f}(d)$

19:   $\mathbb{B}_t \leftarrow \mathbb{B}_{t-1} \bigcup \{m_t\}$

20:   $t \leftarrow t+1$

21: **end while**

22: $\mathbb{M}_{s,N} \leftarrow \{k \in \mathbb{B}_{s,N} \,|\, u_t(d_k) \leq \tau_t\} \forall s \in \mathbb{S}$

23: $\hat{k}_{s,N} \leftarrow \arg\max_{k \in \mathbb{M}_{s,N}} U(\hat{f}_s(d_k), \hat{g}_s(d_k)) \forall s \in \mathbb{S}$

24: **Output:** $\hat{d}_{s,N} \leftarrow d_{\hat{k}_{s,N}} \forall s \in \mathbb{S}$

Overview. We present pseudocode for SAFE-T in Algorithm 1. SAFE-T provides *selection rules* for dose allocation during the course of a clinical trial and a final *recommendation rule* at trial completion for selection of optimal doses

by subgroup for future study. During the trial, allocation occurs in two stages
(with respective selection rules) to better balance safety and learning: a safe
exploration stage, where allocation to unexplored doses is encouraged; and a
Bayesian optimization phase, where allocation is safely optimized with respect
to efficacy. SAFE-T runs for a pre-specified N participants, with the first stage
occurring over a pre-specified N_0 timesteps, where $N_0 < N$. Initial participants
(for each subgroup) are first assigned to the lowest dose, d_1.

SAFE-T models the dose-response functions of toxicity, $\hat{g}_s(d)$, and efficacy,
$\hat{f}_s(d)$ as two separate multi-output Gaussian processes (GP), using the linear
model of co-regionalization [10], with the correlated outputs of each respective
Gaussian process corresponding to each subgroup s. The GPs are trained using
stochastic variational inference [7]. While SAFE-T requires priors set on certain hyperparameters, the use of Gaussian processes eliminates the need for
specifying the explicit parametric form of the dose-responses, as is done in the
commonly used continual reassessment method and other proposed model-based
dose-finding methods [14,20,30]. GPs allow for flexibility of the shape of the
dose-response relationships, which is beneficial as efficacy may take on a monotonically increasing, plateauing, or unimodal shape. We are also able to take
advantage of GP confidence intervals for dose allocation and informative safety
constraints.

At the conclusion of a trial, SAFE-T provides a *recommendation rule* for
determining the optimal dose for each subgroup to be used in future study. This
rule uses a notion of utility concerning dose-finding trials as proposed by [24]
and further used in [12]. It was proposed for use in dose allocation throughout a
trial; however, we use it for both the final dose recommendation rule and post-
hoc performance analysis. It is a weighted L_p norm that evaluates the trade-off
between toxicity and efficacy. We use the slightly modified variation proposed by

[12]: $U(p_E, p_T) = 1 - \left(\left(\frac{p_T}{\tau_T} \right)^p + \left(\frac{1-p_E}{1-\tau_E} \right)^p \right)^{\frac{1}{p}}$; p_E refers to probability of efficacy
and p_T refers to probability of toxicity. The parameter p is set by an elicitation
procedure from clinical trial practitioners (Appendix A.1).

During Trial: Safe Exploration Stage. During the safe exploration stage,
SAFE-T maintains a set of doses that have been previously sampled, $\mathbb{B}_{s,t}$,
for each subgroup s. An available set of doses, $\mathbb{A}_{s,t} = \mathbb{B}_{s,t} \bigcup \{\max(\mathbb{B}_{s,t}) + 1\}$,
includes all doses that have been previously sampled and the next highest
dose (unless the highest dose has been sampled already). $\mathbb{B}_{s,t}$ is empty at the
beginning of the trial; a patient belonging to a subgroup that has not yet
been encountered is assigned to the lowest dose, d_1. The safe set of doses,
$\mathbb{M}_{s,t} = \{k \in \mathbb{A}_{s,t} \,|\, \mu_{\hat{g}_s(d_k)} \leq \tau_t\}$, for each subgroup s, is determined with
respect to the mean of the GP posterior on toxicity for doses in the available
set. SAFE-T allocates the highest safe dose available, if this dose has not been
previously sampled by subgroup s. If all safe doses have been previously sampled, SAFE-T selects the dose with the largest confidence interval on $\hat{g}_s(d)$.
Confidence interval widths are referenced by $c_t(d_k) = 2\beta\sigma_{\hat{g}_s(d_k)}$ for each dose

d_k, where β is a scalar hyperparameter. Thus, dose m_t is selected as follows:

$$m_t = \begin{cases} \max\left(\mathbb{M}_{h_t,t}\right), & \text{if } \max\left(\mathbb{M}_{h_t,t}\right) \notin \mathbb{B}_{h_t,t} \\ \arg\max_{k \in \mathbb{M}_{h_t,t}} c_t(d_k), & \text{otherwise} \end{cases}.$$

During Trial: Safe Optimization Stage. During the rest of the trial, SAFE-T selects the dose from the safe set of doses with the highest expected improvement on efficacy. First, we define the upper bound on the confidence interval for toxicity as $u_t(d_k) = \mu_{\hat{g}_s(d_k)} + \beta\sigma_{\hat{g}_s(d_k)}$. While the available set of doses, $\mathbb{A}_{s,t}$, is defined as in the previous stage, the safe dose set, $\mathbb{M}_{s,t} = \{k \in \mathbb{A}_{s,t} \,|\, u_t(d_k) \leq \tau_t\}$, is defined slightly differently in this stage, now incorporating the confidence intervals on the GP posterior on toxicity. SAFE-T then allocates the dose $m_t = \arg\max_{k \in \mathbb{M}_{h_t,t}} EI(d_k)$, $EI(x)$ referring to expected improvement, a commonly used acquisition function [9].

After Trial: Final Dose Recommendation. At the conclusion of the trial, SAFE-T selects a final recommended dose for each subgroup based on a utility measure, $U(p_e, p_t)$, that incorporates the final posteriors on the GP toxicity and efficacy functions. The safe dose set is composed of all doses that have been allocated during the trial with upper confidence bounds on toxicity below the toxicity threshold: $\mathbb{M}_{s,N} = \{k \in \mathbb{B}_{s,N} \,|\, u_N(d_k) \leq \tau_t\}$. The final recommended dose for each subgroup is the dose with the maximum utility out of the safe dose set: $\hat{k}_{s,N} = \arg\max_{k \in \mathbb{M}_{s,N}} U(\hat{f}_s(d_k), \hat{g}_s(d_k))$; $\hat{d}_{s,N} = d_{\hat{k}_{s,N}}$.

4   Preliminary Results

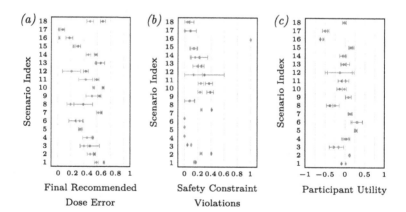

Fig. 1. SAFE-T (orange) consistently outperforms C3T (blue) with (a) lower or similar dose error in all scenarios, (b) fewer or similar number of safety constraint violations in 17/18 scenarios, and (c) higher or similar participant utility in all scenarios. Circular points represent the metric mean across the 2 subgroups, while line caps represent the metric for each subgroup. Further experiment details are shown in Appendix A.2 (Color figure online)

We compare SAFE-T to the C3T algorithm proposed by [14], as it the only related work that also addresses heterogeneous patients in dose-finding trials. We assess performance across 18 synthetic dose-finding scenarios that comprehensively evaluate performance with respect to differing toxicity (monotonically increasing and plateauing) and efficacy (monotonically increasing, plateauing, unimodal) curve shapes and realistic variations between subgroups, as well as possible edge cases. We work with a population of $N = 51$, $N_0 = 18$, with 2 subgroups arriving at $\pi = [0.5, 0.5]$. The reported metrics are averages over 100 trials of each method and averaged across the trial participants. Results are shown in Fig. 1; SAFE-T appears in orange while C3T appears in blue. SAFE-T consistently outperforms C3T in all three categories: final recommended dose error, which assesses whether the recommended dose is equivalent to the true optimal dose; safety constraint violations, which records the number of times a dose with a true toxicity probability greater than the toxicity threshold is allocated to a patient; and participant utility, which is assessed post-hoc based on the dose allocated to each participant. We note that while the utility measure is used to determine final dose recommendations in SAFE-T, it is **not** used during the dose allocation procedure; it is thus a notable result that SAFE-T maintains high utility across many scenarios.

5   Conclusion

In this paper, we present a method for conducting safe dose-finding trials while maintaining high recommended dose accuracy and participant utility. Our algorithm is constructed to be compatible with the realistic constraints of a dose-finding trial, including effectiveness with small sample sizes and heterogeneous participants. Currently, SAFE-T is limited to settings with binary outcome variables and also adheres to common simplifying assumptions, such as that outcomes are observed without delay. Future work could consider more complex scenarios, such as continuous outcomes, delayed outcomes, multiple outcomes, and missing data (as patient dropout can be common in clinical trials).

References

1. Phases of clinical trials (2022). https://www.cancerresearchuk.org/about-cancer/find-a-clinical-trial/what-clinical-trials-are/phases-of-clinical-trials
2. Aziz, M., Kaufmann, E., Riviere, M.K.: On multi-armed bandit designs for dose-finding trials. J. Mach. Learn. Res. **22**, 14:1–14:38 (2020)
3. Baek, J., Farias, V.F.: Fair exploration via axiomatic bargaining. In: Neural Information Processing Systems (2021)
4. Brøgger-Mikkelsen, M., Ali, Z.S., Zibert, J.R., Andersen, A.D., Thomsen, S.F.: Online patient recruitment in clinical trials: systematic review and meta-analysis. J. Med. Internet Res. **22** (2020)
5. Chien, I., Deliu, N., Turner, R.E., Weller, A., Villar, S.S., Kilbertus, N.: Multidisciplinary fairness considerations in machine learning for clinical trials. In: 2022 ACM Conference on Fairness, Accountability, and Transparency (2022)

58   I. Chien et al.

6. Dickmann, L.J., Schutzman, J.L.: Racial and ethnic composition of cancer clinical drug trials: how diverse are we? Oncologist **23**(2), 243–246 (2018)
7. Hensman, J., de G. Matthews, A.G., Ghahramani, Z.: Scalable variational gaussian process classification. In: International Conference on Artificial Intelligence and Statistics (2014)
8. Huang, J., et al.: Sample sizes in dosage investigational clinical trials: a systematic evaluation. Drug Design Dev. Ther. **9**, 305–312 (2015)
9. Jones, D.R., Schonlau, M., Welch, W.J.: Efficient global optimization of expensive black-box functions. J. Global Optim. **13**, 455–492 (1998)
10. Journel, A.G., Huijbregts, C.J.: Mining geostatistics (1976)
11. Kazerouni, A., Ghavamzadeh, M., Abbasi, Y., Roy, B.V.: Conservative contextual linear bandits. In: NIPS (2016)
12. Koopmeiners, J.S., Modiano, J.F.: A Bayesian adaptive phase I–II clinical trial for evaluating efficacy and toxicity with delayed outcomes. Clin. Trials **11**, 38–48 (2014)
13. Kurzrock, R., Lin, C., Wu, T.C., Hobbs, B.P., Pestana, R.C., Hong, D.S.: Moving beyond 3+3: the future of clinical trial design. Am. Soc. Clin. Oncol. Educ. Book. Am. Soc. Clin. Oncol. Ann. Meet. **41**, e133–e144 (2021)
14. Lee, H.S., Shen, C., Jordon, J., van der Schaar, M.: Contextual constrained learning for dose-finding clinical trials. ArXiv abs/2001.02463 (2020)
15. Özdemir, B.C., Gerard, C.L., da Silva, C.E.: Sex and gender differences in anticancer treatment toxicity - a call for revisiting drug dosing in oncology. Endocrinology (2022)
16. Pfohl, S.R., Xu, Y., Foryciarz, A., Ignatiadis, N., Genkins, J.Z., Shah, N.H.: Net benefit, calibration, threshold selection, and training objectives for algorithmic fairness in healthcare. In: 2022 ACM Conference on Fairness, Accountability, and Transparency (2022)
17. Raghavan, M., Slivkins, A., Vaughan, J.W., Wu, Z.S.: The externalities of exploration and how data diversity helps exploitation. In: Annual Conference Computational Learning Theory (2018)
18. Ramamoorthy, A., Kim, H.H., Shah-Williams, E., Zhang, L.: Racial and ethnic differences in drug disposition and response: Review of new molecular entities approved between 2014 and 2019. J. Clin. Pharmacol. **62** (2021)
19. Riviere, M.K., Yuan, Y., Jourdan, J.H., Dubois, F., Zohar, S.: Phase I/II dose-finding design for molecularly targeted agent: plateau determination using adaptive randomization. Stat. Methods Med. Res. **27**, 466–479 (2018)
20. Shen, C., Wang, Z., Villar, S.S., van der Schaar, M.: Learning for dose allocation in adaptive clinical trials with safety constraints. In: International Conference on Machine Learning (2020)
21. Steinberg, J.R., et al.: Analysis of female enrollment and participant sex by burden of disease in us clinical trials between 2000 and 2020. JAMA Netw. Open **4** (2021)
22. Sui, Y., Gotovos, A., Burdick, J.W., Krause, A.: Safe exploration for optimization with gaussian processes. In: International Conference on Machine Learning (2015)
23. Sui, Y., Zhuang, V., Burdick, J.W., Yue, Y.: Stagewise safe Bayesian optimization with gaussian processes. In: International Conference on Machine Learning (2018)
24. Thall, P.F., Cook, J.D.: Dose-finding based on efficacy-toxicity trade-offs. Biometrics **60**, 684–693 (2004)
25. Thomas, M., Bornkamp, B., Seibold, H.: Subgroup identification in dose-finding trials via model-based recursive partitioning. Stat. Med. **37**, 1608–1624 (2018)

26. Unger, J.M., et al.: Sex differences in risk of severe adverse events in patients receiving immunotherapy, targeted therapy, or chemotherapy in cancer clinical trials. J. Clin. Oncol. **40**, 1474–1486 (2022)
27. Villar, S.S., Bowden, J., Wason, J.M.S.: Multi-armed bandit models for the optimal design of clinical trials: benefits and challenges. Stat. Sci.: Rev. J. Inst. Math. Stat. **30**(2), 199–215 (2015)
28. Wages, N.A., Chiuzan, C., Panageas, K.S.: Design considerations for early-phase clinical trials of immune-oncology agents. J. Immunother. Cancer **6**, 1–10 (2018)
29. Wang, Z., Wagenmaker, A.J., Jamieson, K.G.: Best arm identification with safety constraints. ArXiv **abs/2111.12151** (2021)
30. Wheeler, G.M., et al.: How to design a dose-finding study using the continual reassessment method. BMC Med. Res. Methodol. **19**, 1–15 (2019)
31. Zhang, W., Sargent, D.J., Mandrekar, S.J.: An adaptive dose-finding design incorporating both toxicity and efficacy. Stat. Med. **25**, 2365–2383 (2006)
32. Zucker, I., Prendergast, B.J.: Sex differences in pharmacokinetics predict adverse drug reactions in women. Biol. Sex Differ. **11**, 1–14 (2020)

CGXplain: Rule-Based Deep Neural Network Explanations Using Dual Linear Programs

Konstantin Hemker[(✉)], Zohreh Shams, and Mateja Jamnik

Department of Computer Science and Technology, University of Cambridge, Cambridge, UK
{konstantin.hemker,zohreh.shams,mateja.jamnik}@cl.cam.ac.uk

Abstract. Rule-based surrogate models are an effective and interpretable way to approximate a Deep Neural Network's (DNN) decision boundaries, allowing humans to easily understand deep learning models. Current state-of-the-art decompositional methods, which are those that consider the DNN's latent space to extract more exact rule sets, manage to derive rule sets at high accuracy. However, they a) do not guarantee that the surrogate model has learned from the same variables as the DNN (alignment), b) only allow optimising for a single objective, such as accuracy, which can result in excessively large rule sets (complexity), and c) use decision tree algorithms as intermediate models, which can result in different explanations for the same DNN (stability). This paper introduces **C**olumn **G**eneration e**X**plainer to address these limitations – a decompositional method using dual linear programming to extract rules from the hidden representations of the DNN. This approach allows optimising for any number of objectives and empowers users to tweak the explanation model to their needs. We evaluate our results on a wide variety of tasks and show that CGX meets all three criteria, by having exact reproducibility of the explanation model that guarantees stability and reduces the rule set size by >80% (complexity) at improved accuracy and fidelity across tasks (alignment).

Keywords: XAI · Surrogate Models · Deep Learning Explanations · Rule Extraction · Explainability

1 Introduction

In spite of state-of-the-art performance, the opaqueness and lack of explainability of DNNs has impeded their wide adoption in safety-critical domains such as healthcare or clinical decision-making. A promising solution in eXplainable Artificial Intelligence (XAI) research is presented by global rule-based *surrogate models*, that approximate the decision boundaries of a DNN and represent these boundaries in simple IF-THEN-ELSE rules that make it intuitive for humans to interact with [16, 20]. Surrogate models often use *decompositional* approaches, which

© The Author(s), under exclusive license to Springer Nature Switzerland AG 2023
H. Chen and L. Luo (Eds.): TML4H 2023, LNCS 13932, pp. 60–72, 2023.
https://doi.org/10.1007/978-3-031-39539-0_6

inspect the latent space of a DNN (e.g., its gradients) to improve performance, while *pedagogical* approaches only utilise the inputs and outputs of the DNN.

In pursuit of the most accurate surrogate models, recent literature has primarily focussed on improving the fidelity between the DNN and the surrogate model, which refers to the accuracy of the surrogate model when predicting the DNN's outputs $\hat{y}$ instead of the true labels y. While state-of-the-art methods achieve high fidelity [6,8], there are several qualitative problems with these explanations that hinder their usability in practice and have been mostly neglected in previous studies. First, if features are not fully independent, there is no guarantee that a surrogate model has learned from the same variables as the DNN, meaning that the surrogate model may provide misleading explanations that do not reflect the model's behaviour (**alignment**). Second, most rule extraction models optimise for the accuracy of the resulting rule set as a single objective, which can result in excessively large rule sets containing thousands of rules, making them impractical to use (**complexity**). Third, existing decompositional methods use tree induction to extract rules, which tends to be unstable and can result in different explanations for the same DNN, sometimes leading to more confusion than clarification (**stability**).

This paper introduces CGX (Fig. 1) – a flexible rule-based decompositional method to explain DNNs at high alignment and stability, requiring only a fraction of the rules compared to current state-of-the-art methods. We combine and extend two recent innovations of decompositional explanations (i.e., using information from the hidden layers of the DNN) [8] and rule induction literature (i.e., generating boolean rule sets for classification) [7]. First, we suggest a paradigm shift for rule-based surrogate explanations that goes beyond optimising for accuracy as a single objective, allowing users to tailor the explanation to their needs. Concretely, we formulate the objective function of the intermediate model penalises the predictive loss as well as the number of rules and terms as a joint objective. Additionally, CGX allows to easily introduce further objectives. Second, we use a column generation approach as intermediate models, which have proven to be more accurate and stable than tree induction and other rule mining methods. Third, our algorithm introduces *intermediate error prediction*, where the information of the DNN's hidden layers is used to predict the error of the pedagogical solution (Eq. 1). Fourth, we reduce the noise created by adding all rules from the DNN's latent representation by a) conducting direct *layer-wise substitution*, which reduces error propagation of the recursive substitution step used in prior methods and b) dismisses rules that do not improve the performance of the explanation model. This also reduces the need to choose between decompositional and pedagogical methods, since CGX converges to the pedagogical solution in its worst case performance.

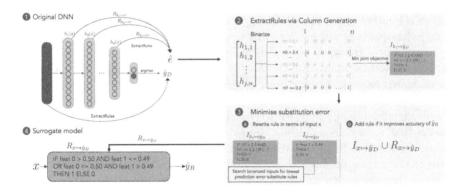

Fig. 1. Overview of the decompositional CGX algorithm, showing the process to get from the DNN as starting point (1) to the explanation model (4) that approximates the DNN's decision boundaries. We 1(a) extract the rule set $R_{x \mapsto \hat{y}_D}$ by training an intermediate model on the DNN's predictions, and 1(b) on the error of that initial ruleset for each hidden layer. Our intermediate extraction through Column generation (2) allows optimising for multiple objectives to extract short and concise rule sets. The substitution step (3) rewrites the intermediate rules $I_{h_j \mapsto \hat{y}_D}$ in terms of the input variables $I_{x \mapsto \hat{y}_D}$ and adds them to the surrogate model (4) if they increase its fidelity.

Contributions

- *Quality metrics*: We formalise three metrics (alignment, complexity, stability) that surrogate explanations need to achieve to be feasibly applied as an explanation model across datasets.
- *Alignment*: We improve alignment between the original and surrogate models, achieving 1–2% higher fidelity of the rule-based predictions and 10–20% higher Ranked Biased Overlap (RBO) of ranked feature importance representations.
- *Complexity*: We reduce the size of the rule sets used to explain the DNN, achieving rule sets with >80% less terms compared to state-of-the-art decompositional baselines.
- *Stability*: Our explanations are guaranteed to produce identical explanations for the same underlying model.
- *Decompositional value*: We demonstrate that decompositional methods are particularly useful for harder tasks, while pedagogical methods are sufficient for simple tasks.

2 Related Work

XAI and Rule-Based Explanations. XAI research has the objective of understanding *why* a machine learning model makes a prediction, as well as *how* the process behind the prediction works [3]. This helps to increase trustworthiness [9], identifying causality [13], as well as establishing confidence, fairness [17], and accessibility [1] in model predictions. *Global explainability methods* attempt to learn a representation that applies to every sample in the data, instead of only individual samples or features (local), and then provide a set of generalisable principles, commonly referred to as a *surrogate model* [3]. Surrogate models can be either pedagogical or decompositional [10]. **Pedagogical methods** train an explainable model on the predictions of the DNN $\hat{y}$ instead of the true labels y, still treating keep treating the DNN as a black-box [5,15]. Pedagogical methods have a faster runtime since they ignore the latent space of the DNN, but sacrifice predictive performance [20]. **Decompositional methods** inspect the model weights or gradients and can therefore learn a closer representation of *how* the model makes a prediction at the expense of runtime.

One promising category of global decompositional methods are rule extraction models such as DeepRED [20], REM-D [16], ECLAIRE [8], and DeXIRE [6]. These methods learn a set of conjunctive (CNF) or disjunctive normal form (DNF) rules $R_{x \mapsto \hat{y}}$ that approximate the neural network's predictions $\hat{y}$ [20]. Existing decompositional methods often use decision tree algorithms, such as C5.0 [14], for intermediate rule extraction. Thus, they learn rules that represent the relationship between each hidden layer and the DNN predictions $R_{h_i \mapsto \hat{y}}$, which are then recursively substituted to be rewritten in terms of the input features as $R_{x \mapsto \hat{y}}$ [16]. While existing surrogate methods achieve high fidelity, the resulting rule set R is often still too large (thousands of rules) to clarify the model's behaviour in practice. Recent research has attempted to reduce the complexity of rule-based surrogates by running different decision tree algorithms, pruning methods [16], or clause-wise substitution [8]. However, existing rule-based surrogate algorithms are heavily dependent on tree-based models used for rule generation. Thus, the performance is significantly sacrificed if the tree depth is too heavily restricted, despite reducing the size of the rule set.

Rule Induction Methods. Another approach to explainability is to use explainable-by-design models, one of which are rule-based representations. Many of these methods use rule mining which first produces a set of candidate terms and then implements a rule selection algorithm which selects or ranks the rules from that search space. The problem with this is that the search space is inherently restricted [11,18]. Another class of methods, such as RIPPER [4] construct their rule sets by greedily adding the conjunction that explains most of the remaining data. This approach comes with the problem that the rule sets are not guaranteed to be globally optimal and commonly result in large rule sets. Two popular state-of-the-art rule induction methods that aim to control rule set complexity are Bayesian Rule Sets (BRS) [18] and Boolean rules from Column Generation (CG). BRS use probabilistic models with prior parameters to construct small-size DNF rule sets. Column generation uses binarisation and large linear programming techniques to efficiently search over the exponential number of possible terms, where the rule set size can be restricted with a complexity constraint in the objective function. While all of the above rule induction methods could be used for the rule extraction, we chose CG due to its stability and flexible formulation of the objective function.

3   Methodology

3.1   Quality Metrics

To improve on the shortcomings of existing decompositional methods, we first provide formal definitions to measure alignment, complexity, and stability. We assume an original model $f(x)$ (DNN) with i hidden layers h_i and the rule-based surrogate model $g(f(x))$ consisting of the rule set $R_{x \mapsto \hat{y}}$ that was extracted using an intermediate model $\psi(\cdot)$.

We define **complexity** as the size of the explanation rule set $|R_{x \mapsto \hat{y}}|$, expressed as the sum of the number of terms of all rules in R, i.e., $\min |R_{x \mapsto \hat{y}}|$.

We measure **alignment** between f_x and g_x in two different ways. First, we look at the **performance alignment** as fidelity, which measures the predictive accuracy of the model predictions $\hat{y}_g$ agains the original model predictions $\hat{y}_f$ as $\mu_{f,g} = \frac{1}{n} \sum_1^{n-1} (\hat{y}_f = \hat{y}_g)$. Second, we assess the **feature alignment** of the resulting explanations. Feature importance is a commonly used to understand which variables a model relies on when making predictions, represented as a ranked list. To validate that f_x and g_x are well-aligned, we want to ensure that both models rely on the same input features from X in their predictions. Assuming two ranked lists S and T, we calculate the Ranked Biased Overlap φ_{ST} [19] as $\max \varphi(S,T,p) = \max(1-p) \sum_{d=1} p^{d-1} A_d$, where A_d is the ratio of list overlap size at depth d and w_d is the geometric progression $w_d = (1-p)p^{d-1}$, a weight vectors used to calculate the weighted sum of all evaluation depths.

Finally, we define **stability** as rule sets that are identical on repeated calls of the explanation methods with the same underlying model. We run the explanation model g_x on different seeds $s = \{0, 1, 2 ..., j\}$, where we want to ensure that the rule sets are equivalent as $R_{x \mapsto \hat{y}}(s_1) = R_{x \mapsto \hat{y}}(s_2)$.

3.2   Column Generation as Intermediate Model

We hypothesise that the majority of the complexity, stability, and alignment issues stem from the choice of the intermediate model $\psi(\cdot)$ in state-of-the-art decompositional methods. We use an adapted version of the column generation solver outlined in [7]. Instead of using $\psi(\cdot)$ as a standalone model, we will show that the column generation solver is well-suited as intermediate model in decompositional methods instead of commonly used tree-based algorithms such as C4.5/C5.0 [16,20]. We start with the original restricted Master Linear Program which formulates from [7] the Hamming loss, which counts the number of terms that have to be removed to classify the incorrect sample correctly. The Hamming loss is bound by an error and complexity constraint. We update the negative reduced cost of the pricing subproblem from [7] to include the hyperparameters for the number of rules (λ_0) and the number of terms (λ_1), which are linked to the complexity constraint as a dual variable. This formulation also makes it simple to add further parameters to the complexity constraint and negative reduced cost (e.g., adding a constraint that penalises rules or terms for only one particular class).

3.3 CG Explainer

ECLAIRE outperforms other decompositional methods on fidelity, rule set size, and run time. Using column generation instead of tree induction as the intermediate model $\psi(\cdot)$, we reformulate the ECLAIRE algorithm as shown in Algorithm 1 with the core objective of improving the three quality metrics we set out. We introduce two versions of the column generation explainer – a pedagogical (CGX-ped) and a decompositional implementation (CGX-dec).

CGX-ped extracts rules from the intermediate model to predict the DNN predictions $\hat{y}_D$. This method ignores the latent space of the DNN, but can still outperform standalone column generation by guidance of the DNN's predictions:

$$\hat{y}_{ped} = R_{x \mapsto \hat{y}_D}(X) = \psi(X, \hat{y}_D) \tag{1}$$

CGX-dec (Algorithm 1) introduces three key innovations over other decompositional methods. First, we do not start with an empty rule set, but uses the pedagogical solution (Eq. 1) as a starting point (line 2). Second, building on the pedagogical rule set, the algorithm iterates through the hidden layers. To improve on the pedagogical solution at each layer, we run *intermediate error prediction* by extracting rules by applying the intermediate model $\psi(\cdot)$ to predict the prediction error of the pedagogical solution $\hat{e}$ from each hidden layer (line 5). That is, we specifically learn rules that discriminate between false and correct prediction of the current best rule set, therefore resulting in rules that would improve this solution. The final update is the substitution method – previous approaches recursively replace the rules [16] or terms [8] of the hidden layer h_{j+1} with the terms for each output class from the previous layer h_j until all hidden rules can be rewritten in terms of the input features X. Since not every hidden layer can be perfectly represented in terms of the input, the substitution step always contains an error which propagates down the layers as the same method is applied recursively. Instead, we use the direct rule substitution step outlined in Algorithm 2. Similar to the CG solver, we first binarise our input features as rule thresholds (line 1). After computing the conjunctions of the candidate rules, we calculate the error for each candidate and select the set of candidate terms with the lowest error (Algorithm 2, line 3) compared to the hidden layer predictions $(y_{\hat{h}_{ij}})$. Knowing that the substitution step still contains an error, some rules contribute more to the performance than others (rules with high errors are likely to decrease predictive performance). Therefore, the last update in Algorithm 1 is that the substituted rules resulting after the substitution step are only added to the rule set if they improve the pedagogical solution (lines 9 & 10).

Algorithm 1 CGX-dec

Input: DNN f_θ with layers $\{h_0, h_1, ..., h_{d+1}\}$
Input: Labelled Training data $X = \{x^{(1)}, ..., x^{(N)}\}; Y = \{y^{(1)}, ..., y^{(N)}\}$
Output: Rule set $R_{x \mapsto \hat{y}}$

1: $\hat{y}^{(1)}, ..., \hat{y}^{(N)} \leftarrow \arg\max(h_{d+1}(x^{(1)})), ..., \arg\max(h_{d+1}(x^{(N)}))$
2: $R_{x \mapsto \hat{y}} \leftarrow \psi(X, \hat{y})$
3: **for** hidden layer $i = 1, ..., d$ **do**
4:     $x'^{(1)}, ..., x'^{(N)} \leftarrow h_i(x^{(1)}), ..., h_i(x^{(N)})$
5:     $\hat{e} \leftarrow (\hat{y}_{ped} \neq \hat{y}_{nn})$
6:     $R_{h_i \mapsto \hat{e}} \leftarrow \psi(\{(x'^{(1)}, \hat{e}_1), ..., (x'^{(N)}, \hat{e}_N)\})$
7:     **for** rule $r \in R_{h_i \mapsto \hat{e}}$ **do**
8:         $s \leftarrow \texttt{substitute}(r)$
9:         $I_{x \mapsto \hat{y}} \leftarrow s \cup R_{x \mapsto \hat{y}}$
10:        **if** $fid(\hat{y}_I, \hat{y}) > fid(\hat{y}_R, \hat{y})$ **then**
11:            $R_{x \mapsto \hat{y}} \leftarrow I_{x \mapsto \hat{y}} \cup R_{x \mapsto \hat{y}}$
12:        **end if**
13:    **end for**
14: **end for**
15: **return** $R_{x \mapsto \hat{y}}$

---

4 Experiments

Given the alignment, complexity, and stability shortcomings of existing methods, we design computational experiments to answer the following **research questions**:

- **Q1.1 Performance alignment:** Does the proven higher performance of column generation rule sets lead to higher fidelity with the DNN?
- **Q1.2 Feature alignment:** How well do aggregate measures such as feature importance from the rule set align with local explanation methods of the DNN?
- **Q2 Complexity:** Can we control the trade-off between explainability (i.e., low complexity) and accuracy by optimising for a joint objective?
- **Q3 Stability:** Do multiple runs of our method produce the same rule set for the same underlying model?

Algorithm 2 Direct rule substitution

Input: rule $r_{h_{ij} \mapsto \hat{y}}$
Input: Training data $X = \{x^{(1)}, ..., x^{(N)}\}$
Hyperparameter: # of rule candidate combinations k
Output: substituted rule(s) $r_{x \mapsto \hat{y}}$

1: $X_{bin} \leftarrow \texttt{BinarizeFeatures}(X, bins)$
2: $r_{cand} \leftarrow \texttt{ComputeConjunctions}(k, X_{bin})$
3: $Errors_{r_{cand}} \leftarrow 1 - \frac{1}{N}\sum^{N-1}(\hat{y}_{h_{ij}} = \hat{y}_{r_{cand}})$
4: $r_{x \mapsto \hat{y}} \leftarrow min(Errors_{r_{cand}})$
5: **return** $r_{x \mapsto \hat{y}}$

- **Q4 Decompositional value:** Is the performance gain of decompositional methods worth the higher time complexity compared to pedagogical methods?

4.1 Baselines and Setup

We use both pedagogical and decompositional explanation baselines in our experiments. For pedagogical baselines, we re-purpose state-of-the-art rule induction and decision tree methods to be trained on the DNN predictions $\hat{y}$ instead of the true labels y. Concretely, we use the C5.0 decision tree algorithm [14], Bayesian Rule Sets [18], and RIPPER [4]. As decompositional baselines, we implement the ECLAIRE algorithm as implemented in [8] which has been shown to outperform other decompositional methods in both speed and accuracy. Additionally, we benchmark against the standalone Column Generation method [7] trained on the true labels y to show the benefit of applying it as an intermediate model in both pedagogical and decompositional settings. We run all baselines and our models on five different real-world and synthetic classification datasets, showing the scalability and adaptability to different numbers of samples, features, and class imbalances.

We run all experiments on five different random folds to initialise the train-test splits of the data, the random initialisations of the DNN as well as random inputs of the baselines. All experiments were run on a 2020 MacBook Pro with a 2GHz Intel i5 processor and 16 GB of RAM. For running the baselines, we use open-source implementations published in conjunction with RIPPER, BRS, and ECLAIRE, running hyperparameter search for best results as set out in the respective papers. For comparability, we use the same DNN topology (number and depth of layers) as used in the experiments in [8]. For hyperparameter optimisation of the DNN, we use the `keras` implementation of the Hyperband algorithm [12] to search for the optimal learning rate, hidden and output layer activations, batch normalisation, dropout, and L2 regularisation. The `CGX` implementation uses the MOSEK solver [2] as its `cvxpy` backend. The code implementation of the `CGX` algorithm can be found at https://github.com/konst-int-i/cgx.

5   Results

Performance Alignment (Q1.1). The primary objective of performance alignment is the fidelity between the predictions of the rule set $\hat{y}_R$ compared to the model predictions $\hat{y}_{DNN}$, since we want an explanation model that mimics the DNNs behaviour as closely as possible. The results in Table 1 show that `CGX-ped` has a higher fidelity compared to the baseline methods on most datasets by approximately 1-2% whilst having significantly fewer rules. While RIPPER has a slightly higher fidelity on the MAGIC dataset, both `CGX-ped` and `CGX-dec` achieve competitive performance whilst only requiring 5% of the rules. This table also shows that a high fidelity does not guarantee a high accuracy on the overall task, which is visible on the FICO dataset. While `CGX` achieves a very high fidelity in this task, the overall accuracy is relatively low. This is caused by the

Table 1. Overview of `CGX-ped` and `CGX-dec` performance alignment (fidelity) and complexity (# terms) compared to the baselines across datasets. `CGX-ped` outperforms all baselines across the majority of tasks. While RIPPER has a slightly higher fidelity on the MAGIC dataset, `CGX` only requires ~5% of the terms.

Dataset	Model	Rule Fid.	Rule Acc.	# Rules	# Terms
XOR	CG (STANDALONE)	78.0 ± 16.8	81.1 ± 18.5	5.2 ± 1.9	21.6 ± 12.7
	Ripper (PED)	53.5 ± 3.9	53.8 ± 4.0	7.4 ± 3.6	14.4 ± 7.5
	BRS (PED)	91.3 ± 2.0	95.5 ± 1.3	9.0 ± 0.3	80.9 ± 3.0
	C5 (PED)	53.0 ± 0.2	52.6 ± 0.2	1 ± 0	1 ± 0
	ECLAIRE (DEC)	91.4 ± 2.4	91.8 ± 2.6	87 ± 16.2	263 ± 49.1
	CGX-ped (OURS)	**92.4 ± 1.1**	**96.7 ± 1.7**	3.6 ± 1.8	10.4 ± 7.2
	CGX-dec (OURS)	**92.4 ± 1.1**	**96.7 ± 1.7**	3.6 ± 1.8	10.4 ± 7.2
MAGIC	CG (STANDALONE)	85.7 ± 2.5	82.7 ± 0.3	5.2 ± 0.8	13.0 ± 2.4
	Ripper (PED)	**91.9 ± 0.9**	81.7 ± 0.5	152.2 ± 14.6	462.8 ± 53.5
	BRS (PED)	84.6 ± 2.1	79.3 ± 1.3	5.8 ± 0.3	24.1 ± 4.8
	C5 (PED)	85.4 ± 2.5	82.8 ± 0.9	57.8 ± 4.5	208.7 ± 37.6
	ECLAIRE (DEC)	87.4 ± 1.2	84.6 ± 0.5	392.2 ± 73.9	1513.4 ± 317.8
	CGX-ped (OURS)	90.4 ± 1.7	80.6 ± 0.6	**5.0 ± 0.7**	**11.6 ± 1.9**
	CGX-dec (OURS)	91.5 ± 1.3	84.4 ± 0.8	7.4 ± 0.8	11.6 ± 1.9
MB-ER	CG (STANDALONE)	92.1 ± 1.1	92.0 ± 1.1	5.0 ± 0.7	15.4 ± 2.2
	Ripper (PED)	86.5 ± 2.2	85.2 ± 3.0	22.0 ± 9.2	30.2 ± 21.6
	BRS (PED)	90.9 ± 1.2	88.4 ± 0.9	8.9 ± 1.1	57.6 ± 18.5
	C5 (PED)	89.3 ± 1	92.7 ± 0.9	21.8 ± 3	72.4 ± 14.5
	ECLAIRE (DEC)	**94.7 ± 0.2**	**94.1 ± 1.6**	48.3 ± 15.3	137.6 ± 24.7
	CGX-ped (OURS)	93.7 ± 1.1	92.0 ± 0.9	**4.2 ± 0.4**	**17.0 ± 1.9**
	CGX-dec (OURS)	**94.7 ± 0.9**	92.4 ± 0.7	5.9 ± 1.1	21.8 ± 3.4
MB-HIST	CG (STANDALONE)	88.5 ± 2.3	91.1 ± 1.4	4.0 ± 0.7	19.4 ± 2.4
	Ripper (PED)	86.7 ± 3.7	88.1 ± 3.3	13.8 ± 3.4	35.0 ± 11.6
	BRS (PED)	81.7 ± 2.1	79.9 ± 2.5	5.1 ± 0.2	40.3 ± 5.8
	C5 (PED)	89.3 ± 1	87.9 ± 0.9	12.8 ± 3.1	35.2 ± 11.3
	ECLAIRE (DEC)	89.4 ± 1.8	88.9 ± 2.3	30 ± 12.4	74.7 ± 15.7
	CGX-ped (OURS)	89.1 ± 3.6	89.4 ± 2.5	**5.2 ± 1.9**	**27.8 ± 7.6**
	CGX-dec (OURS)	**89.6 ± 3.6**	**90.2 ± 2.5**	6.8 ± 2.0	32.2 ± 8.3
FICO	CG (STANDALONE)	86.4 ± 2.8	70.6 ± 0.4	3.3 ± 1.1	8.6 ± 3.6
	Ripper (PED)	88.8 ± 2.8	70.2 ± 1.0	99.2 ± 14.5	307.4 ± 41.6
	BRS (PED)	84.8 ± 2.3	65.4 ± 2.1	3.1 ± 0.2	18 ± 3.2
	C5 (PED)	72.7 ± 2.1	81.8 ± 1.6	34.8 ± 4.1	125.6 ± 35.2
	ECLAIRE (DEC)	66.5 ± 2.5	**84.9 ± 1.7**	161.0 ± 12.3	298.0 ± 21.2
	CGX-ped (OURS)	91.1 ± 0.1	70.5 ± 0.8	**3.6 ± 1.1**	**9.6 ± 3.6**
	CGX-dec (OURS)	**92.4 ± 0.2**	71.4 ± 1	5.1 ± 1.3	13.4 ± 2.1

underlying DNN struggling to perform well in this task. Notably, the performance of `CGX-dec` and `CGX-ped` is equivalent on the XOR dataset, indicating that there were no rules to add from the intermediate layers. This is because the XOR dataset is a relatively simple synthetic dataset, where the pedagogical version already identifies nearly the exact thresholds that were used to generate the target (see Fig. 2b).

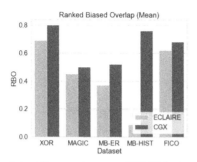

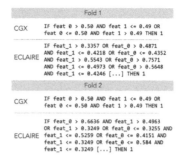

(a) Alignment scores (RBO) between   (b) Ruleset Stability across folds
     CGX-dec and ECLAIRE

Fig. 2. Overview of CGX performance with respect to alignment (a) and stability (b). Subfigure (a) shows the mean Ranked Biased Overlap of CGX compared to ECLAIRE shows that CGX's rule set show a higher *feature alignment*. Subfigure (b) shows that CGX has exact reproducibility for the same underlying model.

Feature Alignment (Q1.2). Going beyond fidelity, the feature alignment score ψ in Fig. 2a shows the mean RBO score ψ between the feature importance derived from the CGX rule set and the aggregated importance of local methods (SHAP and LIME) of the original DNN. A higher score shows that the two ranked lists are more aligned and, as such, the DNN and the rule-based surrogate model rely on the same features for their explanations more closely. Figure 2a compares the decompositional CGX-dec method to the best-performing decompositional baseline (ECLAIRE) and shows that CGX-dec achieves a higher feature alignment across all datasets compared to the baseline.

Complexity (Q2). Table 1 shows that both the pedagogical and decompositional methods achieve highly competitive results with only a fraction of the rules required. Compared to pedagogical baselines, CGX-ped outperforms on the majority of the tasks. While the pedagogical BRS baseline produces fewer rules for some datasets (FICO and MB-HIST), their total number of terms are more than double those of CGX across all datasets due to longer chained rules being produced by this method. Additionally, the BRS fidelity is not competitive with CGX-ped or CGX-dec. Looking at ECLAIRE as our decompositional baseline, the results show that CGX-dec only requires a fraction of the terms compared to ECLAIRE. In the case of the Magic dataset, ECLAIRE required >100x more rules than our method, while for other datasets, the multiple ranges from 10–20x more rules required.

Stability (Q3). Figure 2b shows that CGX (both versions) results in identical explanations when running only the explainability method on a different random seed, keeping the data folds and random seed of the DNN identical. We observe that CGX produces the exact same rule set on repeated runs, while our decompositional baseline produces different explanations, which can be confusing to users. Note that this stability is different from the standard deviation shown in Table 1, where we would expect variation from different splits of the data and random initialisations of the DNN.

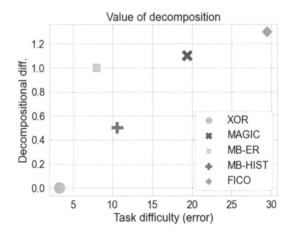

Fig. 3. Comparison of task difficulty (DNN prediction error, x-axis) with the incremental fidelity improvement (y-axis) when using `CGX-dec` over `CGX-ped`. As tasks get more difficult, using `CGX-dec` adds relatively more fidelity compared to `CGX-ped`.

Value of Decomposition (Q4). We acknowledge that the time complexity of decompositional methods scales linearly to the number of layers, which makes the pedagogical `CGX-ped` implementation an attractive alternative for very deep network topologies. To help decide whether to use pedagogical or decompositional methods, we looked at how much the information from the DNN's latent space (lines 3–10 in Algorithm 1) improves the pedagogical solution (line 2 in Algorithm 1). Figure 3 shows that the added performance gained from information of the hidden layers is related to the difficulty of the task. For "easy" tasks, (i.e., those where the DNN has a high accuracy/AUC such as the XOR task), `CGX-ped` and `CGX-dec` converge to the same solution, since no rules from the hidden layers increase the fidelity. Figure 3 shows that the performance difference increases with the difficulty of the task. For the FICO task, where the DNN accuracy is only just over 70%, the surrogate model gains the most information from the hidden layers.

6 Discussion

This paper introduces a global decompositional method that uses column generation as intermediate models. We improve rule-based explanations by intermediate error predictions from the latent space of a DNN, coupled with layer-wise substitution to reduce error propagation. `CGX` enables research and industry to customise surrogate explanations for different end users by parameterising the accuracy-explainability trade-off. First, we introduced a quantitative measure to analyse the **feature alignment** between the surrogate model and local explanations of the DNN and show that our surrogate model explanations are more closely aligned to other local explanation methods of the original model. Second, the design of the objective functions allows assigning a higher cost to surrogate model complexity (i.e., the number of terms in the rule set) using an extra hyper-parameter. We demonstrate that this achieves significantly **lower complexity** and enables users to control the accuracy-interpretability trade-off by setting higher or lower penalties on the number of rules. Third, the results show that CGX is independent of its initialisation (solution to the Master Linear Program), which leads to **improved stability** compared to methods using tree induction for rule extraction. Additionally, CGX requires **fewer hyperparameters** compared to tree-based algorithms such as C5, hence requiring less fine-tuning to achieve competitive results. While this introduces the lambda parameter to enable users to control the length of the resulting rule set, it is also possible to run the solver unconstrained. Beyond these benefits, having rule-based surrogate models enables end users to *intervenability* by users, as they can amend the rule set to encode further domain knowledge.

The key limitation of CGX and decompositional methods more generally is that the runtime is highly dependent on the number of hidden DNN layers and the number of columns in X. We attempt to mitigate this problem by showing that CGX-ped is a highly competitive alternative, especially for simple tasks. For more difficult tasks, however, the decompositional method still delivers better explanations with higher fidelity. The implementation will be open-sourced as a pip-installable Python package.

Acknowledgments. KH acknowledges support from the Gates Cambridge Trust via the Gates Cambridge Scholarship.

References

1. Adadi, A., Berrada, M.: Peeking inside the black-box: a survey on explainable artificial intelligence (XAI). IEEE Access **6**, 52138–52160 (2018)
2. Andersen, E.D., Andersen, K.D.: The MOSEK interior point optimizer for linear programming: an implementation of the homogeneous algorithm. In: Frenk, H., Roos, K., Terlaky, T., Zhang, S. (eds.) High Performance Optimization. Applied Optimization, vol. 33, pp. 197–232. Springer, Boston (2000). https://doi.org/10.1007/978-1-4757-3216-0_8
3. Arrieta, A.B., et al.: Explainable artificial intelligence (XAI): concepts, taxonomies, opportunities and challenges toward responsible AI. Inf. Fusion **58**, 82–115 (2020)
4. Cohen, W.W.: Fast effective rule induction. In: Machine Learning Proceedings 1995, pp. 115–123 (1995). https://doi.org/10.1016/B978-1-55860-377-6.50023-2
5. Confalonieri, R., Weyde, T., Besold, T.R., Moscoso del Prado Martín, F.: Trepan reloaded: a knowledge-driven approach to explaining artificial neural networks (2020)
6. Contreras, V., et al.: A dexire for extracting propositional rules from neural networks via binarization. Electron. (Switz.) **11** (2022). https://doi.org/10.3390/ELECTRONICS11244171
7. Dash, S., Gunluk, O., Wei, D.: Boolean decision rules via column generation. Adv. Neural Inf. Process. Syst. **31** (2018)
8. Espinosa, M., Shams, Z., Jamnik, M.: Efficient decompositional rule extraction for deep neural networks. In: XAI Debugging Workshop @ NEURIPS 2021 (2021)

9. Floridi, L.: Establishing the rules for building trustworthy AI. Nat. Mach. Intell. **1**(6), 261–262 (2019)

10. Islam, S.R., Eberle, W., Ghafoor, S.K., Ahmed, M.: Explainable artificial intelligence approaches: a survey. arXiv preprint arXiv:2101.09429 (2021)

11. Lakkaraju, H., Bach, S.H., Leskovec, J.: Interpretable decision sets: a joint framework for description and prediction. In: Proceedings of the 22nd ACM SIGKDD International Conference on Knowledge Discovery and Data Mining, pp. 1675–1684 (2016)

12. Li, L., Jamieson, K., Rostamizadeh, A., Talwalkar, A.: Hyperband: a novel bandit-based approach to hyperparameter optimization. J. Mach. Learn. Res. **18**, 1–52 (2018). http://jmlr.org/papers/v18/16-558.html

13. Murdoch, W.J., Singh, C., Kumbier, K., Abbasi-Asl, R., Yu, B.: Interpretable machine learning: definitions, methods, and applications. arXiv preprint arXiv:1901.04592 (2019)

14. Pandya, R., Pandya, J.: C5. 0 algorithm to improved decision tree with feature selection and reduced error pruning. Int. J. Comput. Appl. **117**(16), 18–21 (2015)

15. Saad, E.W., Wunsch, D.C., II.: Neural network explanation using inversion. Neural Netw. **20**(1), 78–93 (2007)

16. Shams, Z., et al.: REM: an integrative rule extraction methodology for explainable data analysis in healthcare. bioRxiv (2021)

17. Theodorou, A., Wortham, R.H., Bryson, J.J.: Designing and implementing transparency for real time inspection of autonomous robots. Connect. Sci. **29**(3), 230–241 (2017)

18. Wang, T., Rudin, C., Doshi-Velez, F., Liu, Y., Klampfl, E., MacNeille, P.: A Bayesian framework for learning rule sets for interpretable classification. J. Mach. Learn. Res. **18**(1), 2357–2393 (2017)

19. Webber, W., Moffat, A., Zobel, J.: A similarity measure for indefinite rankings. ACM Trans. Inf. Syst. (TOIS) **28**(4), 1–38 (2010)

20. Zilke, J.R., Loza Mencía, E., Janssen, F.: DeepRED – rule extraction from deep neural networks. In: Calders, T., Ceci, M., Malerba, D. (eds.) DS 2016. LNCS (LNAI), vol. 9956, pp. 457–473. Springer, Cham (2016). https://doi.org/10.1007/978-3-319-46307-0_29

ExBEHRT: Extended Transformer for Electronic Health Records

Maurice Rupp$^{(\boxtimes)}$, Oriane Peter, and Thirupathi Pattipaka

Novartis Oncology AG, Basel, Switzerland
maurice.rupp@gmail.com , thirupathi.pattipaka@novartis.com

Abstract. In this study, we introduce ExBEHRT, an extended version of BEHRT (BERT applied to electronic health record data) and applied various algorithms to interpret its results. While BEHRT only considers diagnoses and patient age, we extend the feature space to several multi-modal records, namely demographics, clinical characteristics, vital signs, smoking status, diagnoses, procedures, medications and lab tests by applying a novel method to unify the frequencies and temporal dimensions of the different features. We show that additional features significantly improve model performance for various down-stream tasks in different diseases. To ensure robustness, we interpret the model predictions using an adaption of expected gradients, which has not been applied to transformers with EHR data so far and provides more granular interpretations than previous approaches such as feature and token importances. Furthermore, by clustering the models' representations of oncology patients, we show that the model has implicit understanding of the disease and is able to classify patients with same cancer type into different risk groups. Given the additional features and interpretability, ExBEHRT can help making informed decisions about disease progressions, diagnoses and risk factors of various diseases.

Keywords: BERT · RWE · Patient Subtyping · Interpretability

1 Introduction

Over the last decade, electronic health records have become increasingly popular to document a patient's treatments, lab results, vital signs, etc. Commonly, a sequence of medical events is referred to as a *patient journey*. Given the immense amount of longitudinal data available, there lies tremendous potential for machine learning to provide novel insights about the recognition of disease patterns, progression and subgroups as well as treatment planning. Recent studies have adapted transformers to structured tabular EHR data and demonstrated their superiority in various benchmarks compared to other similar algorithms ([4]). Although there exists work on unstructured freetext EHR data (e.g. BioBERT ([5])), these models are out of the scope of this study.

© The Author(s), under exclusive license to Springer Nature Switzerland AG 2023
H. Chen and L. Luo (Eds.): TML4H 2023, LNCS 13932, pp. 73–84, 2023.
https://doi.org/10.1007/978-3-031-39539-0_7

2   Related Work

The first adaptation of transformers to structured EHR data, called BEHRT ([7]), incorporated diagnosis concepts and age from EHRs and added embeddings for the separation of individual visits and a positional embedding for the visit number. Other models such as Med-BERT ([14]), CEHR-BERT ([10]) and BRLTM ([9]) added more features by concatenating the inputs into one long patient sequence. These approaches are limited in the amount of data from a single patient they can process and the computational power required increases significantly with each feature added. In addition, there exists a variety of models that either combine the BERT architecture with other machine learning models ([6,11,15]) or focus only on disease-specific use-cases ([1,12,13]). These models lack generalizability to other tasks due to their unique training methodologies and domains.

In this work, we present a novel approach to integrate multimodal features into transformer models by adding medical concepts separately and vertically instead of chaining all concepts horizontally. We show that these features are important in various downstream applications such as mortality prediction, patient subtyping and disease progression prediction.

3   ExBEHRT for EHR Representation Learning

ExBEHRT is an extension of BEHRT where medical concepts are not concatenated into one long vector, but grouped into separate, learnable embeddings per concept type. In this way, we avoid exploding input lengths when adding new medical features and give the model the opportunity to learn which concepts it should focus on. From a clinical perspective, it would also be stringent to separate diagnoses, procedures, drugs, etc., as they have different clinical value for downstream applications. We take the number of diagnoses in a visit as an indicator of how many "horizontal slots" are available for other concepts in that visit (e.g. two for the first visit in Fig. 1). Therefore, the maximum length of the patient journey is defined by the number of diagnosis codes of a patient, regardless of the number of other concepts added to the model. As shown by the procedures in Fig. 1, but carried out in the same way with lab tests, there are three possible cases of adding a new concept to a visit:

a) The number of procedures is equal to the amount of horizontal slots available in the visit (visit 1 - two each). The procedures can therefore be represented as a 1D vector.

b) The number of procedures exceeds the amount of slots available in the visit (visit 2 - one diagnosis, two procedures). Here, the procedures fill up the number of horizontal slots line by line until there are no more procedures left, resulting in a 2D vector of dimensions $\#slots \times \lceil \frac{\#procedures}{\#slots} \rceil$.

c) The number of procedures subceeds the amount of slots available (visit 3 - one diagnosis, no procedures). The procedures are represented as a 1D vector and then padded to the amount of horizontal slots available.

After reshaping, all procedures and labs of all patients are padded to the same amount of rows n to enable batch processing. Before passing the inputs to the model, each token is embedded into a 288-dimensional vector and all tokens are summed vertically. Figure 7 in the appendix shows the final representation of one patient.

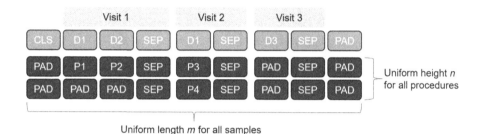

Fig. 1. An example of how ExBEHRT represents a sample patient. As the features are stacked vertically, additional concepts (such as labs as shown in Fig. 7) will not increase the sentence length m.

3.1 Data

In this study, we used the Optum® de-identified EHR database. It is derived from healthcare provider organizations in the United States, which include more than 57 contributing sources and 111,000 sites of care including hospital-based medical services networks comprising academic, private, and community hospitals treating more than 106 million patients. Optum® data elements also include demographics, medications prescribed and administered, immunizations, allergies, lab results (including microbiology), vital signs and other observable measurements, clinical and hospitalisation administrative data, and coded diagnoses and procedures. The population in Optum® EHR is geographically diverse, spanning all 50US states. We selected only data points collected during hospitalisations to ensure data quality and consistency. Each patient must have at least five visits with valid ICD-9 or ICD-10 diagnosis codes to ensure sufficient temporal context. Considering these criteria, our final pre-training cohort consisted of 5.4 million individual patients divided into training (80%), validation (10%) and testing (10%) groups. Table 1 shows the characteristics of the final cohort.

Table 1. Statistics of the pre-training cohort.

Feature	Metric
Birth year	1973 ± 25, min: 1932, max: 2021
Gender	41.49% male, 58.51% female
Distribution by race	68% Cau., 22% Afr. Am., 1% As., 9% other
No. of diagnosis codes per patient	14 ± 11.1, min: 5, max: 121
No. of visits per patient	9 ± 6.6, min: 5, max: 63
% of patients without labs	14.33%
% of patients without procedures	1.64%
% of patients without BMI	21.74%
% of patients without smoking status	27.11%
% of deceased patients	14.52%

3.2 Model Training

ExBEHRT consists of the same model architecture as BEHRT. For pre-training, we applied the standard MLM procedure described in the original BERT paper for predicting masked diagnosis codes using their BertAdam optimizer with cross-entropy loss. All BERT-based models (ExBEHRT, BEHRT, Med-BERT) were trained for 40 epochs on one Tesla T4 GPU with 16GB memory, selecting the epoch with the highest micro-averaged MLM precision score on the validation set. To ensure a fair comparison, we used the same amount of attention layers (6) and heads (12) as well as embedding dimension (288) for all three BERT-based models. We also pretrained a version of ExBEHRT on the additional pre-training objective PLOS[1] as introduced by Med-BERT, which we called ExBEHRT+P.

In a second step, we fine-tuned the models on two prediction tasks: Death of a patient within six months after the first cancer diagnosis and readmission into hospital within 30 or fewer days after heart failure. All tokens after the cancer diagnosis/heart failure are not disclosed to the model. The cohorts for these two tasks were split into 80% training, and 10% each validation and testing datasets, with each patient present in both cohorts (pre-training and fine-tuning) assigned to the same split for both tasks. The cohorts consist of 437'902 patients (31.67% deceased within 6 months after first cancer diagnosis) for Death in 6M and 503'161 patients (28.24% readmitted within 30 days) for HF readmit. Furthermore, we used the patient representations of ExBEHRT to identify risk subtypes of cancer patients using unsupervised clustering. For this purpose, we used a combination of the dimensionality reduction technique UMAP ([8]) and the clustering algorithm HDBSCAN ([2]) and applied the clustering to all cancer patients from the pretraining cohort (260'645).

---

[1] Binary classification of whether a patient had at least one prolonged length of stay in hospital (> 7 days) during their journey.

4 Results

4.1 Event Prediction

For each experiment, we selected the model with the best validation precision score and only then evaluated the performance on the test set. The metrics used for evaluation are the area under the receiver operating characteristic curve (AUROC), average precision score (APS) as well as the precision at the 0.5 threshold. In all but one metric in one task, ExBEHRT outperforms BEHRT, MedBERT and other conventional algorithms such as Logistic Regression (LR) and XGBoost when evaluated on this hold-out dataset (Table 2).

Table 2. Average fine-tuning results of various models and their standard deviations.

Task	Metric	LR	XGB	BEHRT	Med-BERT	ExBEHRT	ExBEHRT+P
Death in 6M	APS	42.8 ± 0.0%	45.5 ± 0.1%	47.7 ± 0.4%	46.2 ± 0.4%	**53.1±0.3%**	52.6 ± 0.3%
	AUROC	63.5 ± 0.0%	66.4 ± 0.1%	66.7 ± 0.6%	65.3 ± 0.3%	**71.5 ± 0.5%**	70.9 ± 0.5%
	Precision	73.0 ± 0.1%	74.3 ± 0.1%	75.2 ± 0.2%	74.5 ± 0.1%	**78.1 ± 0.1%**	77.9 ± 0.1%
Death in 12M	APS	51.6 ± 0.0%	45.5 ± 0.1%	55.5 ± 0.1%	54.4 ± 0.2%	**59.8 ± 0.2%**	59.6 ± 0.2%
	AUROC	66.7 ± 0.0%	66.3 ± 0.1%	70.1 ± 0.2%	68.9 ± 0.3%	**74.3 ± 0.4%**	73.8 ± 0.4%
	Precision	70.4 ± 0.1%	74.4 ± 0.1%	73.2 ± 0.1%	72.4 ± 0.1%	**76.4 ± 0.1%**	76.3 ± 0.1%
HF readmit	APS	29.8 ± 0.0%	**31.3 ± 0.1%**	19.9 ± 0.1%	19.8 ± 0.1%	30.0 ± 1.6%	25.1 ± 0.1%
	AUROC	51.9 ± 0.1%	53.6 ± 0.1%	51.2 ± 0.1%	51.0 ± 0.1%	56.7 ± 1.7%	**56.8 ± 0.2%**
	Precision	72.0 ± 0.0%	72.3 ± 0.1%	81.0 ± 0.1%	81.0 ± 0.0%	78.7 ± 0.2%	**81.6 ± 0.1%**

4.2 Interpretability on Event Prediction Results

For all interpretability experiments, we used the ExBEHRT model fine-tuned on the task *Death in 6M*, meaning whether a cancer patient will decease within six months after their first cancer diagnosis. We visualize the interpretability for individual patients only, as both interpretability approaches presented here are example-based and not model-agnostic.

Self-Attention Visualization. Analogous to previous work ([7,9,14]), we visualised the attention of the last network layer using BertViz ([16]). However, since in all such models all embeddings are summed before being passed through the network, self-attention has no way of assigning individual input features to the outcome. Nevertheless, we can draw conclusions about how the different slots interact with each other and which connections the model considers important. Figure 2 shows the self-attention of a single patient in the last layer of ExBEHRT. The left figure shows the attentions of all 12 attention heads in this layer, while the right figure shows the attention of one head. As expected, the model focuses strongly on the slots within a visit, as these slots are by definition strongly interconnected. Slot 7 corresponds to the slot in which the patient was diagnosed with lung cancer. Although the model was not specifically trained on cancer codes, it pays close attention to this slot, indicating that it has learned some correlation between the cancer diagnosis and the predicted outcome. Interestingly, slot 7 receives a lot of attention on the first and second visits, but not on the two previous visits, suggesting that the model is able to learn causality over long time intervals.

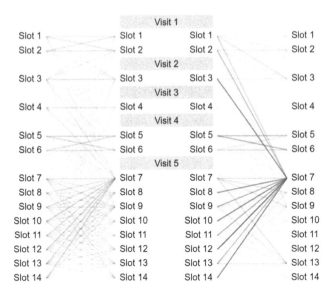

Fig. 2. Left: The self-attention of all 12 attention heads of the last layer of ExBEHRT. Higher opacity corresponds to higher attention. Right: The self-attention of one attention head of the last layer.

Expected Gradients Interpretability. Due to the limitations of self-attention visualisation, we have explored the technique Expected Gradients ([3]) for more detailed interpretability. With this algorithm, we can infer the meaning of individual input tokens, which is not possible with self-attention. Since each token (diagnosis code, procedure code, age, etc.) is mapped to a 288-dimensional embedding before being passed to the model, we first calculated the expected gradients for the embedding and then summed the absolute values to obtain a single gradient value for each token. In this way, each individual token has an associated gradient that is linked to the output of the model and provides detailed insights into which medical concept has what impact on the prediction of the model. Our example patient is a 58-year-old woman who was a regular smoker. She died at the age of 65, three months after her blood cancer diagnosis. In Fig. 3, we summed all expected gradients for each of the input features. This way, we can evaluate the feature importances on the output for a specific patient. For this patient, diagnoses and procedures (treatments & medications) were by far the most importance features. With this visualization, we can further evaluate basic biases. For example, gender was not considered to be an important feature, indicating that predictions would be similar for a person with another gender.

In Fig. 4, we visualized the absolute expected gradients for each of the features and summed them at each time slot. This way, we can evaluate the different feature importances over time to get a notion of where the model puts emphasis on. Interestingly, the model put more importance on what kind of medications

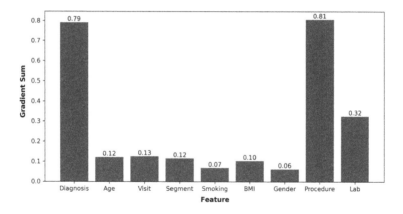

Fig. 3. The absolute sums of the expected gradients summed by input feature.

& treatments that patient received in the first two visits, where as in the last visit (the visit in which the patient was diagnosed with blood cancer), it put more importance on diagnoses and labs. Generally, slot 5, where the cancer was diagnosed, was attributed with the highest importance.

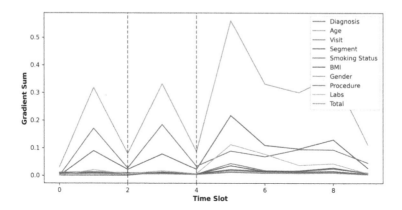

Fig. 4. The absolute sums of the expected gradients summed by input feature and time slot. The dotted lines indicate the next visit.

Fig. 5 displays the absolute sums of gradients of each individual input token, providing a detailed interpretation of which medical concept has had what impact on the models prediction. Unsurprisingly, the cancer code C81 has had the biggest impact on the outcome. However, earlier codes like J40 or 71020 also contribute to the models prediction, indicating that the model includes information from the whole patient journey into its predictions.

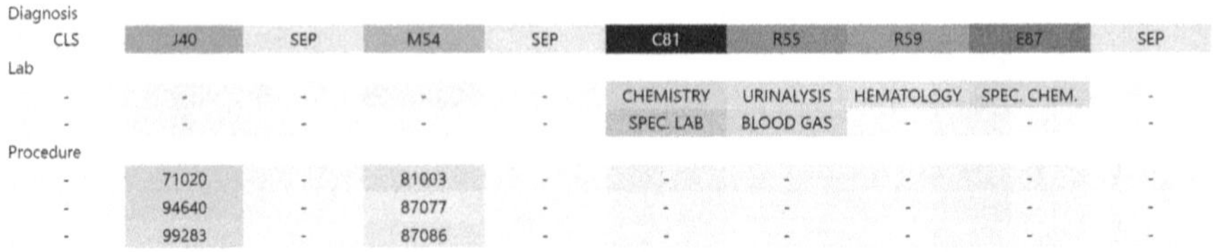

Diagnosis CLS	J40	SEP	M54	SEP	C81	R55	R59	E87	SEP
Lab	-	-	-	-	CHEMISTRY	URINALYSIS	HEMATOLOGY	SPEC. CHEM.	-
	-	-	-	-	SPEC. LAB	BLOOD GAS	-	-	-
Procedure	-	71020	-	81003	-	-	-	-	-
	-	94640	-	87077	-	-	-	-	-
	-	99283	-	87086	-	-	-	-	-

Fig. 5. A visualization of the absolute sums of the expected gradients of diagnoses, labs and procedures on a concept level. Darker colours represent higher values and the SEP tokens indicate the separation between two visits.

4.3   Cancer Patient Clustering

HDBSCAN was able to cluster 90% of all cancer patients into 24 clusters[2]. On average, the most occurring cancer diagnosis within a cluster was present for 84% of the patients assigned to this specific cluster and the mean cluster purity was 85%. Similar concepts (e.g. cancer of female reproductive organs or different types of leukaemia) lay in areas close to each other, indicating a spatial logic between the cancer types. In Fig. 6, we show that with a second pass of HDBSCAN on a given cluster, we can identify risk subgroups. In all three identified clusters, more than 90% of the patients actually do have pancreatic cancer and all clusters share similar general characteristics. However, as shown in the Table

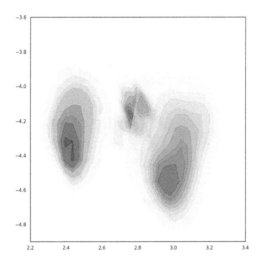

Fig. 6. The three identified patient subclusters with pancreatic cancer visualized with a kernel density estimate plot for visual clarity. Even though the three clusters generally share the same characteristics in diagnoses, Age, BMI etc., patients belonging to the smaller purple cluster died less frequently and recovered nearly twice as often from cancer compared to the other two clusters.

---

[2] A visualization of all clusters can be found in Fig. 8 in the appendix.

Table 3. Statistics of the three pancreatic cancer clusters indicating a clear differentiation between higher risk (gray, blue) and lower risk patients (purple).

Metric	Gray	Blue	Purple
Median age	67	68	68
Median birth year	1950	1947	1944
Median BMI	25	25	26
% of men	52.3%	50.9%	60.0%
Average death rate	76.5%	75.9%	**70.0%**
% of journey with cancer	27.0%	24.0%	**18.3%**
Cancer-free	34.0%	36.9%	**62.7%**

3, ExBEHRT identified a subgroup with a significantly higher chance of recovering from cancer and a lower probability of dying, although this information was not provided to the model at any point in time[3].

5   Discussion

In this study, we presented a novel method for adding patient features to BEHRT that significantly increases the predictive power for multiple downstream tasks in different disease domains. The novel method of stacking features vertically yielded improvements in hardware requirements and benchmarks and facilitates the extension to new concepts in the future. Given the large number and heterogeneity of patients with which the model has been pre-trained, we are confident that ExBEHRT will generalise well to new data, patients and tasks. Combined with interpretability, the model provides more detailed insights into disease trajectories and subtypes of different patients than previous approaches, and could help clinicians form more detailed assessments of their patients' health. Furthermore, with a personalised understanding of patient groups, it is possible to identify unmet needs and improve patient outcomes.

Limitations
Nevertheless, there are some limitations: It is extremely difficult to validate the quality, completeness and correctness of EHR datasets, as the data is usually processed anonymously and comes from a variety of heterogeneous, fragmented sources. The sheer nature of EHR data also introduces bias, as physicians may have an incentive to diagnose additional less relevant conditions, as medical billing is closely related to the number and type of diagnoses reported.

---

[3] In the table, *% of journey with cancer* indicates the ratio of the time between the first and last cancer diagnosis compared to the duration of the whole recorded patient journey. *Cancer-free* refers to the percentage of patients within a cluster, which have records of at least two visits without cancer diagnosis after the last visit with a cancer diagnosis. The *average death rate* comes directly from the EHR database and unfortunately does not include information on the cause of death.

82   M. Rupp et al.

Future Work

In a potential next step, we would like to verify the results and interpretations of this work with clinicians to ensure robust and sound predictions as possible given the acquired interpretability. In addition, we would like to test the generalisability of ExBEHRT to other clinical use-cases such as severity prediction and risktyping of other diseases as well as specific cancers.

Appendix

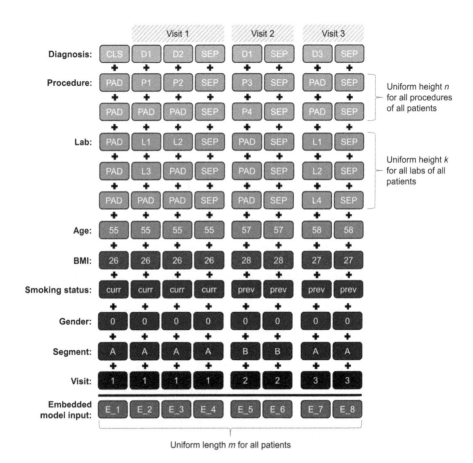

Fig. 7. A sample input of ExBEHRT. Each of the concepts has its own embedding, where each of the tokens is mapped to a 288-dimensional vector, which is learned during model training. After embedding, all concepts are summed vertically element-wise to create a single $288 \times m$ dimensional vector as input for the model.

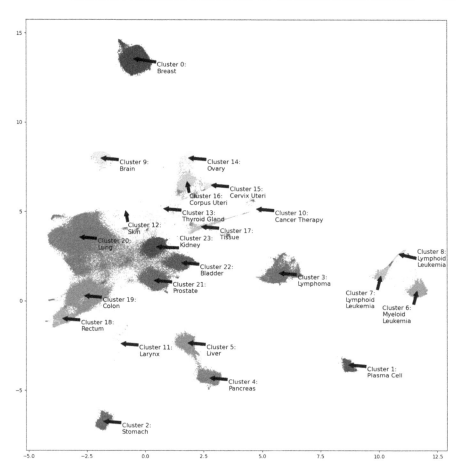

Fig. 8. The unsupervised cluster assignments from HDBSCAN, visualized with a 2-dimensional UMAP projection. The gray points are patients not assigned to any cluster (10%). The labels indicate the most frequent diagnosis code of each cluster. Besides cluster 10, all labels are neoplasms. (Color figure online)

References

1. Azhir, A., et al.: Behrtday: Dynamic mortality risk prediction using time-variant COVID-19 patient specific trajectories. In: AMIA Annual Symposium Proceedings (2022)
2. Campello, R.J.G.B., Moulavi, D., Sander, J.: Density-based clustering based on hierarchical density estimates. In: Pei, J., Tseng, V.S., Cao, L., Motoda, H., Xu, G. (eds.) PAKDD 2013. LNCS (LNAI), vol. 7819, pp. 160–172. Springer, Heidelberg (2013). https://doi.org/10.1007/978-3-642-37456-2_14
3. Erion, G., Janizek, J.D., Sturmfels, P., Lundberg, S.M., Lee, S.I., Allen, P.G.: Improving performance of deep learning models with axiomatic attribution priors and expected gradients. Nature **3**, 620–631 (2020)

4. Kalyan, K.S., Rajasekharan, A., Sangeetha, S.: AMMU: a survey of transformer-based biomedical pretrained language models. J. Biomed. Inf. **126**, 103982 (2022)
5. Lee, J., et al.: BioBERT: a pre-trained biomedical language representation model for biomedical text mining. Bioinformatics **36**, 1234–1240 (2019)
6. Li, Y., et al.: Hi-BEHRT: hierarchical transformer-based model for accurate prediction of clinical events using multimodal longitudinal electronic health records. J. Biomed. Health Inf. **27**, 1106–1117 (2021)
7. Li, Y., et al.: BEHRT: transformer for electronic health records. Nature (2020)
8. McInnes, L., Healy, J., Melville, J.: Umap: Uniform manifold approximation and projection for dimension reduction. J. Open Source Softw. (2018)
9. Meng, Y., Speier, W., Ong, M.K., Arnold, C.W.: Bidirectional representation learning from transformers using multimodal electronic health record data to predict depression. J. Biomed. Health Inf. **25**, 3121–3129 (2021)
10. Pang, C., et al.: CEHR-BERT: incorporating temporal information from structured EHR data to improve prediction tasks. In: Proceedings of Machine Learning for Health (2021)
11. Poulain, R., Gupta, M., Beheshti, R.: Few-shot learning with semi-supervised transformers for electronic health records. In: Proceedings of Machine Learning Research, vol. 182 (2022)
12. Prakash, P., Chilukuri, S., Ranade, N., Viswanathan, S.: RareBERT: transformer architecture for rare disease patient identification using administrative claims. In: Proceedings of the AAAI Conference on Artificial Intelligence (2021)
13. Rao, S., et al.: An explainable transformer-based deep learning model for the prediction of incident heart failure. IEEE J. Biomed. Health Inf. **26**, 3362–3372 (2022). https://doi.org/10.1109/JBHI.2022.3148820
14. Rasmy, L., Xiang, Y., Xie, Z., Tao, C., Zhi, D.: Med-BERT: pre-trained contextualized embeddings on large-scale structured electronic health records for disease prediction. Nature **4**, 86 (2021)
15. Shang, J., Ma, T., Xiao, C., Sun, J.: Pre-training of graph augmented transformers for medication recommendation. Int. Joint Conf. Artif. Intell. (2019)
16. Vig, J.: A multiscale visualization of attention in the transformer model. In: ACL (2019)

STASIS: Reinforcement Learning Simulators for Human-Centric Real-World Environments

Georgios Efstathiadis[1]([✉])([iD]), Patrick Emedom-Nnamdi[1],
Arinbjörn Kolbeinsson[2], Jukka-Pekka Onnela[1], and Junwei Lu[1]

[1] Department of Biostatistics, Harvard T.H. Chan School of Public Health Boston,
Boston, MA 02115, USA
{gefstathiadis,patrickemedom,onnela,junweilu}@hsph.harvard.edu
[2] Evidation Health, London, UK
arinbjorn@evidation.com

Abstract. We present on-going work toward building *Stasis*, a suite of reinforcement learning (RL) environments that aim to maintain realism for human-centric agents operating in real-world settings. Through representation learning and alignment with real-world offline data, Stasis allows for the evaluation of RL algorithms in offline environments with adjustable characteristics, such as observability, heterogeneity and levels of missing data. We aim to introduce environments the encourage training RL agents that are capable of maintaining a level of performance and robustness comparable to agents trained in real-world online environments, while avoiding the high cost and risks associated with making mistakes during online training. We provide examples of two environments that will be part of Stasis and discuss its implications for the deployment of RL-based systems in sensitive and high-risk areas of application.

Keywords: reinforcement learning · health care · real-world simulators

1 Introduction

Reinforcement Learning (RL) is becoming increasingly popular for a variety of tasks, ranging from robotic control and autonomous driving to artificial intelligence in the gaming domain. Despite its potential, the lack of realistic simulators for RL agents operating in the real-world is a major limitation for the development of reliable agents. Current simulators lack the capability to model real-world applications of RL. This includes missing key components such as accounting for heterogeneity within the environment (specifically within the reward function) and observability, as all real-world environment are inherently perceived as partially observable. Furthermore, these simulators often lack the ability to handle missing data, either irregularly sampled data or observations missing at random (due design of the data collection tool used or observed features) and missing not-at-random (due to outcomes).

H. Chen and L. Luo (Eds.): TML4H 2023, LNCS 13932, pp. 85–92, 2023.
https://doi.org/10.1007/978-3-031-39539-0_8

Lastly, they lack the ability to generate observed data, as simulators should be thought of as a generative model of the real-world, where we want to generate samples close to the observed data.

The lack of realistic simulators for RL agents hinders the development of agents that can be successfully deployed to real-world tasks. The high cost and risk associated with inaccurate predictions during online training makes it an important problem to address. To this end, we introduce Stasis[1], a suite of RL environments that aim to maintain realism for human-centric agents operating in real-world environments. Through representation learning and alignment with real-world offline data, Stasis allows RL systems to be trained in offline environments with tunable characteristics, such as observability, heterogeneity and levels of missing data. The resulting RL agents are capable of maintaining a level of performance and robustness that is comparable to agents trained in real-world online environments, while avoiding the high cost and risk associated with making mistakes during online training.

Related Work. The most similar work to the one we present here is the Gymnasium, formerly known as OpenAI gym as seen in [5], and the Safety Gym by [21]. Both of these are suites of environments where RL agents can be trained without requiring real-world deployment. However, they both place emphasis on robotics and control, with Safety Gym making use of MuJoCo [23] with a focus on constrained RL. The Stasis library which we introduce here will focus on open problems related to RL in healthcare, including partial observability, heterogeneity, missing data and make use of labelled real-world data through offline RL [11].

2 Underlying Framework and Considerations

On-going efforts to build simulated environments for benchmarking conventional and emerging RL algorithms center on emulating the realism and practicality of real-world settings. Evaluating algorithms in this fashion affords practitioners the ability to rigorously examine the suitability of an algorithm before initial deployment in the real world. In this paper, we identify four *core* themes that are important for representing *human-centric* real-world settings. Specifically, settings where the decision-making policy directly interacts with a human, or provides actions for a human to execute within their own environment.

Observability. Observability determines the full-range of information from the environment available to the agent for decision-making. In real-world settings, environments are typically partially observable; the agent only has access to a limited view of the current state of the environment [12]. This can make it difficult to learn an optimal policy, as the agent may be missing important information or have to rely on incomplete observations to determine its actions. Therefore, a well-designed observability mechanism that captures the relevant

---

[1] We plan on releasing the code for Stasis and for the two environments later this year.

information is critical for learning a good policy. However, increasing observability can also lead to higher computational and memory requirements, making
it important to strike a balance between having enough information to make
informed decisions and keeping the complexity manageable.

Heterogeneity. In real-world settings, the reward signal may vary between
agents operating within a single environment. As such, learning a single policy
that aims to optimize the reward for all agents is often difficult, leading to
sub-optimal performance for select agents [6], [8]. Generally, this can result in
a situation where some agents learn different, unintended behaviors. In multi-
agent systems, this can lead to a lack of coordination, potentially hampering the
functioning of the overall system. In applications such as healthcare where data
from heterogeneous subjects are often used to make decisions for single subject,
failing to account for heterogeneity in the reward signal can lead to an alignment
problem, severely impacting the relevancy of the learned policy. Mitigating these
challenges may require algorithms to leverage techniques from areas of research
such as multi-agent reinforcement learning, or to directly modify the reward
functions to account for the heterogeneity between agents.

Missing Data. The effectiveness of a policy in reinforcement learning is closely
tied to the quality and quantity of data used to train the model. If the agent
encounters missing data, such as incomplete or unavailable state or reward information, it may be unable to accurately estimate the value of different actions,
leading to suboptimal decisions [2,13,22]. Missing data can happen due to irregular sampling, where data is missing at random, which can occur due to various
factors such as technical failures or data collection constraints. Additionally, data
may be missing not at random, such as when specific actions or states are more
likely to be absent. In healthcare application, this can be due to phenomena
such as self-selection bias, where participants in the study exercise control over
whether or not they participate in the study or how much data they provide.
Therefore, it is essential to consider the consequences of missing data and address
it using techniques such as imputation, data-augmentation, or other advanced
methods for handling missing data in reinforcement learning [2,22].

Offline Data. Previously collected experiential data from agents interacting
within a given environment can be used to enhance the robustness and reliability of the simulated environment. We envision that offline data can be used to
improve the following aspects of the simulated environment: (1) *state representation* – offline data can be used to provide more accurate state representations
for the agent, including information about the environment, objects, and other
agents [24]; (2) *model dynamics* – the interactions between objects and agents
in the environment can be modeled more accurately using offline data, allowing
for a more realistic representation of the environment's dynamics [9]. Lastly, in
most real-world environments, decision-making policies are rarely trained from
scratch, rather offline data is commonly used to learn policies that achieve an
acceptable level of performance [11]. As such, incorporating available offline data

into simulated environments allows for a pre-training phase, where policies are first initialized using offline data before being deployed within the environment.

3   The Simulator

The structure of the simulator is similar to the structure of the Gymnasium API [5]. An environment is pulled from the library's collection and then any type of agent can be trained using the simulated environment. Each environment has the same method structure, in order for the users to be able to switch among environments and train on different scenarios with ease.

The difference to the Gymnasium API is that the environments also share parameters related to problems found in real world applications, in order to make the environments more realistic and thus the agents more robust to real world data. When initializing an environment, the complexity of the problem will be specified, but also some parameters that are important in a healthcare context [3] and which are problematic in the collection and curation of healthcare data [20]. The shared parameters, when it is possible for an environment, will be able to tune aspects such as the heterogeneity of the simulation [1], incorporate missing data or have partial observability and add stochasticity or noise to the simulation. This can look different for every environment, but the purpose of the parameters is shared.

3.1   Healthy Traveling Salesman

The first environment is a simulated weighted travelling salesman problem [14]. Studies have shown that certain environmental exposures are associated with healthcare bio-markers, e.g. greenspace exposure is associated with lower levels of depression [10]. The problem this environment represents is finding the optimal routes to maximize or minimize a certain exposure related to the health of an individual. The environment is initialized by providing a set of coordinates, each of which has to be visited once, and an exposure type (e.g. greenspace or bars). Then, the environment will collect information on the possible routes that can be taken to visit each coordinate using the OpenRouteService API [16] and the exposures around the locations of interest using the Overpass API [18] which both use data collected from OpenStreetMap or OSM for short [19]. In the current implementation, one of the coordinates to be provided is the starting location and the rest of the coordinates are the ones that are visited, with the agent ultimately returning to the starting point. The task is thus finding the optimized circle in a graph, with complexity of the problem being increased by simply adding more coordinates.

The reward for each action is a weighted average between the distance covered and the time spent at exposure at each route, which is also tunable at input depending on what the agent should focus on optimizing. The possible exposure information is limited only by the possible types of locations that are collected from OSM. In terms of the parameters mentioned before on making the simulations more realistic to collected data, possible concepts discussed include modifying heterogeneity by adding different constraints in the possible actions of different simulated users. Some users have trouble moving large distances or want to avoid certain trigger areas, which can be encompassed in the reward function. In terms of missing data and stochasticity we can tune the amount of information and noise we see in the possible routes. They can also be encompassed in a way that matches what we see from GPS collecting devices like smartphones and smartwatches, where missing data are not missing at random, but there are certain time-periods for which data are not being collected by the smart devices [4].

The following Figs. 1 and 2 are examples of rendering of the environment, where the green areas are greenspace locations [17] collected by the Overpass API, the blue arrows are the coordinates of the locations to be visited and the red home arrow indicates the coordinates of the starting and ending location. This is the visualization after an episode has been run using a Deep-QN agent [15] trained on maximizing greenspace exposure on a set of 8 coordinates in Boston, MA (Fig. 1) and a set of 6 coordinates in Bronx, NY (Fig. 2).

Fig. 1. Map Environment rendering (Boston, MA)

Fig. 2. Map Environment rendering (Bronx, NY)

Researchers that want to use this environment and possess offline GPS data can also encompass them to enrich the information in the environment and make it even more realistic [7]. Using GPS trajectories, information can be collected on areas that people want to avoid or areas with more traffic and this information is reflected in the reward function of the environment. The GPS data can also be used to gain information of people's home and work locations or locations they like to visit frequently, thus making the environment adjust to a specific person's patterns and visit locations.

3.2 Resource Allocation in Clinical Settings

The second environment model in the Stasis library demonstrates a common problem encountered in clinical settings: dynamic resource allocation. This environment's properties can be highly complex due to the sophistication of modern clinical settings. In order to efficiently and operationally allocate resources, decision-making must be carried out on a case-by-case basis, considering the resources available, the individual conditions of multiple patients, and the associated costs and durations of the resources in question.

This environment's properties can be highly complex due to the sophistication of modern clinical settings. However, for its first iteration, it will be limited to a general setting. The state space includes the set of available resources and their characteristics, the occupancy of the clinical section, the time and date, other features that help forecast future occupancy, and relevant patient features, outcomes of utilized resources, and further diagnosis. The action space involves selecting resources from a given available set, which can be adjusted through domain expertise to incorporate best practices. The main goal of this framework is to understand the relationship between resources and patient outcomes and allow the agent to explore different strategies in the safe, non-destructive environment of the simulator.

4   Future Directions

As the initiative grows, it will be important to focus on community building. This can be accomplished by creating a leaderboard, hosting workshops, and adding existing standalone environments. Another focus of the library will be taking advantage of existing data to build more realistic environments. By leveraging existing offline data, the library could potentially use algorithms such as pretraining or initialization phases to further refine the environment and help it to behave in the most realistic way possible. Finally, there should be an active goal to make the environments relevant and useful in a medical or clinical context. To do this, researchers and developers will seek to collaborate with medical professionals to ensure the simulators are based on real world observations and are as accurate as possible. By doing so, Stasis can become a valuable tool for medical professionals.

References

1. Angus, D.C., Chang, C.C.H.: Heterogeneity of treatment effect. JAMA **326**(22), 2312 (2021). https://doi.org/10.1001/jama.2021.20552
2. Awan, S.E., Bennamoun, M., Sohel, F., Sanfilippo, F., Dwivedi, G.: A reinforcement learning-based approach for imputing missing data. Neural Comput. Appl. **34**(12), 9701–9716 (6 2022). https://doi.org/10.1007/S00521-022-06958-3/TABLES/13, https://doi.org/10.1007/s00521-022-06958-3
3. Awrahman, B.J., Aziz Fatah, C., Hamaamin, M.Y.: A review of the role and challenges of big data in healthcare informatics and analytics. Comput. Intell. Neurosci. **2022**, 1–10 (2022). https://doi.org/10.1155/2022/5317760
4. Barnett, I., Onnela, J.P.: Inferring mobility measures from GPS traces with missing data. Biostatistics **21**(2), e98–e112 (2018). https://doi.org/10.1093/biostatistics/kxy059

5. Brockman, G., et al.: OpenAI Gym. arXiv: Learning (2016). https://www.arxiv.org/pdf/1606.01540
6. Chen, E.Y., Song, R., Jordan, M.I.: Reinforcement learning with heterogeneous data: estimation and inference (2022). https://doi.org/10.48550/arxiv.2202.00088, https://arxiv.org/abs/2202.00088v1
7. Gur, I., Nachum, O., Faust, A.: Targeted environment design from offline data (2022). https://openreview.net/forum?id=Is5Hpwg2R-h
8. Jin, H., Peng, Y., Yang, W., Wang, S., Zhang, Z.: Federated reinforcement learning with environment heterogeneity (2022). https://proceedings.mlr.press/v151/jin22a.html
9. Kidambi, R., Rajeswaran, A., Netrapalli, P., Joachims, T.: MOReL: model-based offline reinforcement learning. In: Advances in Neural Information Processing Systems 2020-Decem (2020). https://doi.org/10.48550/arxiv.2005.05951, https://arxiv.org/abs/2005.05951v3
10. Klein, Y., Lindfors, P., Osika, W., Hanson, L.L.M., Stenfors, C.U.: Residential greenspace is associated with lower levels of depressive and burnout symptoms, and higher levels of life satisfaction: a nationwide population-based study in Sweden. Int. J. Environ. Res. Public Health **19**(9), 5668 (2022). https://doi.org/10.3390/ijerph19095668, https://www.mdpi.com/1660-4601/19/9/5668/pdf?version=1651915974
11. Levine, S., Kumar, A., Tucker, G., Fu, J.: Offline reinforcement learning: Tutorial, review, and perspectives on open problems. arXiv preprint arXiv:2005.01643 (2020)
12. Littman, M.L.: A tutorial on partially observable Markov decision processes. J. Math. Psychol. **53**(3), 119–125 (2009). https://doi.org/10.1016/J.JMP.2009.01.005
13. Lizotte, D.J., Gunter, L., Laber, E.B., Murphy, S.A.: Missing data and uncertainty in batch reinforcement learning (2008)
14. Lu, H., Zhang, X., Yang, S.: A learning-based iterative method for solving vehicle routing problems. In: International Conference on Learning Representations (2020). https://www.openreview.net/pdf?id=BJe1334YDH
15. Mnih, V., et al.: Playing Atari with deep reinforcement learning. arXiv: Learning (2013). http://cs.nyu.edu/koray/publis/mnih-atari-2013.pdf
16. Neis, P., Zipf, A.: OpenRouteService.org - combining open standards and open geodata. The state of the map. In: 2nd Open Street Maps Conference, Limerik. Ireland (2008)
17. Novack, T., Wang, Z., Zipf, A.: A system for generating customized pleasant pedestrian routes based on OpenStreetMap data. Sensors **18**(11), 3794 (2018). https://doi.org/10.3390/s18113794
18. Olbricht, R.M.: Data retrieval for small spatial regions in OpenStreetMap. In: Jokar Arsanjani, J., Zipf, A., Mooney, P., Helbich, M. (eds.) OpenStreetMap in GIScience. LNGC, pp. 101–122. Springer, Cham (2015). https://doi.org/10.1007/978-3-319-14280-7_6
19. OpenStreetMap contributors: Planet dump retrieved from (2017). https://planet.osm.org, https://www.openstreetmap.org
20. Pezoulas, V.C., et al.: Medical data quality assessment: on the development of an automated framework for medical data curation. Comput. Biol. Med. **107**, 270–283 (2019). https://doi.org/10.1016/j.compbiomed.2019.03.001
21. Ray, A., Achiam, J., Amodei, D.: Benchmarking safe exploration in deep reinforcement learning. arXiv preprint arXiv:1910.01708 **7**(1), 2 (2019)
22. Shortreed, S.M., et al.: Informing sequential clinical decision-making through reinforcement learning: an empirical study. Mach. Learn. **84**, 109–136 (2011). https://doi.org/10.1007/s10994-010-5229-0

23. Todorov, E., Erez, T., Tassa, Y.: MuJoCo: a physics engine for model-based control. In: 2012 IEEE/RSJ International Conference on Intelligent Robots and Systems, pp. 5026–5033. IEEE (2012)
24. Zang, H., et al.: Behavior prior representation learning for offline reinforcement learning (2022). https://doi.org/10.48550/arxiv.2211.00863, https://arxiv.org/abs/2211.00863v2

Cross-Domain Microscopy Cell Counting By Disentangled Transfer Learning

Zuhui Wang$^{(\boxtimes)}$ (iD)

State University of New York at Stony Brook, Stony Brook, NY 11794, USA
zuwang@cs.stonybrook.edu

Abstract. Microscopy images from different imaging conditions, organs, and tissues often have numerous cells with various shapes on a range of backgrounds. As a result, designing a deep learning model to count cells in a source domain becomes precarious when transferring them to a new target domain. To address this issue, manual annotation costs are typically the norm when training deep learning-based cell counting models across different domains. In this paper, we propose a cross-domain cell counting approach that requires only weak human annotation efforts. Initially, we implement a cell counting network that disentangles domain-specific knowledge from domain-agnostic knowledge in cell images, where they pertain to the creation of domain style images and cell density maps, respectively. We then devise an image synthesis technique capable of generating massive synthetic images founded on a few target-domain images that have been labeled. Finally, we use a public dataset consisting of synthetic cells as the source domain, where no manual annotation cost is present, to train our cell counting network; subsequently, we transfer only the domain-agnostic knowledge to a new target domain of real cell images. By progressively refining the trained model using synthesized target-domain images and several real annotated ones, our proposed cross-domain cell counting method achieves good performance compared to state-of-the-art techniques that rely on fully annotated training images in the target domain. We evaluated the efficacy of our cross-domain approach on two target domain datasets of actual microscopy cells, demonstrating the feasibility of requiring annotations on only a few images in a new domain.

Keywords: Cell counting · Knowledge disentangling · Transfer learning

1 Introduction

Counting cells in microscopy images is useful for many biology discoveries and medical diagnoses [14,17,21]. The number of red blood cells and white blood cells in human bone marrow is a critical indicator of blood cell disorder-related diseases, such as Thalassemia and Lymphoma. Microscopy cell counting is also

H. Chen and L. Luo (Eds.): TML4H 2023, LNCS 13932, pp. 93–105, 2023.
https://doi.org/10.1007/978-3-031-39539-0_9

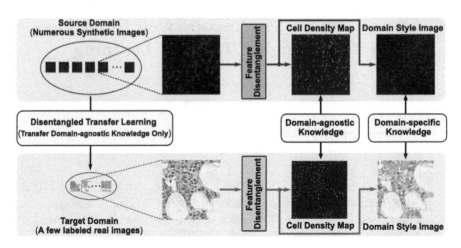

Fig. 1. The challenges and our proposal for the cross-domain cell counting task.

important in drug discovery for assessing the effects of drugs on cellular proliferation and death. These techniques are employed in cell-based assays to predict drug response. Furthermore, cell count numbers are commonly used as a measure of toxicity in high-content screening for small molecules [4]. Counting numbers of crowded cells is a tedious, time-consuming, and error-prone task in real-world applications. Therefore, various deep learning methods have been developed for exact cell counting [5,7,18–20], which is usually achieved by generating a cell density map for a microscopy cell image and then integrating the density map to estimate the total cell number in the image. Three challenges are remaining unsolved in the cell counting problem: (1). **Information entanglement**: a cell image embeds two types of entangled features for cell counting: cell densities in images; and various cell appearances, shapes and various image background contexts related to specific tissues and imaging conditions. Disentangling the mixed features in images and learning informative and discriminative features will facilitate deep learning networks to generate accurate cell density maps while being invariant to specific cells or background contexts; (2). **Costly annotation**: the success of deep learning on cell counting relies on well-annotated training datasets. To relieve the tedious and time-consuming annotation efforts, it is expected to invent weak-annotation cell counting approaches which can learn from a small amount of annotated data and use a large amount of synthetic data; (3). **Large cross-domain gaps**. Transferring a model trained on a source domain (e.g., a synthetic dataset) to a new target domain is a promising approach to adapting a cell counting network to different application scenarios. But, when the data distributions between the source and target domains exhibit large variations, the domain gap will hinder the transfer.

1.1   Our Observations and Proposal

Firstly, observing the information entanglement challenge, we propose to decouple an input cell image into a *cell density map* corresponding to *domain-agnostic* knowledge and a *domain style image* corresponding to *domain-specific* knowledge, as shown in Fig. 1 (right). The domain-agnostic knowledge is related to cell density map generation and remains consistent among various cell counting scenarios, while the domain-specific knowledge related to specific imaging conditions, organs, and tissues is disentangled from the input without affecting the cell counting. Secondly, to reduce the dense annotation cost, we propose to train a cell counting network using a synthetic dataset with known ground truth without manual annotation and then transfer the network to a target domain (Fig. 1 (left)). To address the issue of insufficient training data in the target domain, we propose an image synthesis method that can generate a large amount of training data based on a few annotated target-domain images. Thirdly, the data distribution gaps between domains are mainly caused by domain-specific knowledge that is not shared among various domains, so transferring domain-specific knowledge will have a negative impact on adapting cell counting networks across domains. Thus, during transfer learning, we propose to only transfer disentangled domain-agnostic knowledge that is consistent among domains. The contributions of this paper are threefold:

- A novel cell counting network is designed to estimate cell numbers by disentangling input cell images into domain-agnostic knowledge (cell density maps) and domain-specific knowledge (domain style images);
- A new image synthesis method is proposed which synthesizes cell image patches based on a few real images in the target domain and blends a random number of cells into random locations in domain style images, yielding a large number of training images in the target domain for transfer learning;
- A new progressive disentangled transfer learning is proposed for cross-domain cell counting, which effectively transfers the informative and discriminative features learned from a synthetic source-domain dataset to a new target domain with weak annotated images.

2   Related Work

Cell counting methods can be broadly classified into two categories: detection-based methods [1,2] and regression-based methods [5,9,11,20]. Detection-based methods use a detector to locate cells in images, and the total number of cells is then estimated by counting the detected cells. However, the accuracy of these methods heavily depends on the detector's performance, which can be challenging due to occlusions, various cell shapes, and complex background environments in microscopy images. On the other hand, regression-based methods have recently emerged as state-of-the-art models for cell counting [9]. These methods estimate the total number of cells directly using regression models, without the

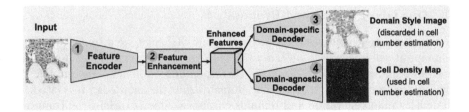

Fig. 2. Our proposed cell counting network architecture with domain-agnostic and domain-specific knowledge disentangled.

need for explicit detection. Given their promising performance, this paper primarily focuses on discussing and comparing several regression-based microscopy cell counting methods.

Most of the state-of-the-art deep learning networks for regression-based cell counting employ density map generation techniques. For instance, [20] proposed an FCN-based model that generates density maps to estimate cell numbers directly. Similarly, SAU-Net [7] is an improved model that incorporates a self-attention module into U-Net [13] to enhance cell counting performance. Another example is Count-ception [5], which estimates cell numbers using specially designed cell density maps generated using square kernels. However, these models require domain-specific training data with dense annotations in various domains to achieve promising cell counting results. The availability of a sufficient number of annotated samples is crucial for the success of these state-of-the-art cell counting algorithms.

3   Methodology

In this section, we first introduce a cell counting network which is capable of disentangling domain-agnostic/specific features. Then, we describe how to synthesize target-domain images from a few weak annotated samples. Finally, we present how to transfer a cell counting network trained on a synthetic source domain to a target domain by preserving discriminative information.

3.1   Cell Counter Aware of Domain-Agnostic/Specific Knowledge

Cell Counter Architecture. The proposed cell counter model is illustrated in Fig. 2 with four main modules: (1) feature encoder; (2) feature enhancement module; (3) domain-specific decoder; and (4) domain-agnostic decoder. First, for the feature encoder, we employ the first ten layers of a pre-trained VGG16 [15] model, which consists of ten convolutional layers with 3×3 kernels. After every two or three convolutional layers, the max-pooling layer is applied. The encoder is intended to extract basic feature representations of input images.

Second, the extracted features are then fed into the feature enhancement module that contains an attention submodule and a dilated convolution submodule [10]. The attention submodule consists of spatial and channel-wise attention mechanisms. The spatial attention contains two convolutional layers, with a sigmoid function followed. This spatial attention tends to focus on important regions within encoded features. The channel-wise attention consists of a global average pooling layer and two dense layers followed by a sigmoid function. The channel-wise attention assigns different weights on feature channels to emphasize contributions of various feature maps. The weighted features are then passed to six dilated convolutional layers to extract features at different scales. The dilated convolutional layer is capable to capture crowded target features (i.e., cell features in our project) at multiple-scale environments [10].

Finally, the enhanced features are sent to the domain-specific decoder and domain-agnostic decoder. Both of these decoders have three up-sampling layers followed by three convolutional layers. The domain-specific decoder attempts to extract features unique to biological experiment domains as domain style images. Simultaneously, the domain-agnostic decoder preserve discriminative features and generates cell density maps for cell number estimation. The domain-specific knowledge, represented by domain style images, varies across domains, so disentangling it out of the input image will enable the transfer learning to correctly transfer the domain-agnostic knowledge shared across domains which control the cell density map generation.

Loss Function. There are two terms for the total loss function (L): Pixel-wise Mean Squared Error (L_{MSE}) loss for the domain-agnostic decoder to generate density maps (i.e., 4th module in Fig. 2), and Perceptual loss (L_{PERC}) for the domain-specific decoder (i.e., 3rd module in Fig. 2) to measure the high-level perceptual and semantic differences [8]. The proposed loss function is written as,

$$L = L_{\text{MSE}} + L_{\text{PERC}} = \frac{1}{N}\sum_{n=1}^{N}\left\|\hat{\mathbf{Y}}_n - \mathbf{Y}_n\right\|_2^2 + \frac{1}{N}\sum_{n=1}^{N}\left\|\hat{\mathbf{\Phi}}_n - \mathbf{\Phi}_n\right\|_2^2, \qquad (1)$$

where $\hat{\mathbf{Y}}_n$ is the n-th generated density map by the domain-agnostic decoder, $\mathbf{Y}_n$ is the corresponding ground truth cell density map. $\hat{\mathbf{\Phi}}_n$ is the feature map of the n-th generated domain style image, $\mathbf{\Phi}_n$ is the corresponding feature map of the ground truth domain style image. N is the total number of samples in the training set.

3.2 Synthesizing Target-Domain Images by a Few Annotated Ones

In this paper, we assume that only a few annotated target-domain images are available, which are insufficient to fine-tune a pre-trained model. Therefore, we consider synthesizing more target-domain images to help a pre-trained model transfer domain-agnostic knowledge across domains. The general workflow of our image synthesis is summarized in Fig. 3 with four modules:

- Module-(a) (Fig. 3(a)): given an annotated image in the target domain, all cell patches with a resolution of 32×32 are cropped based on the annotated

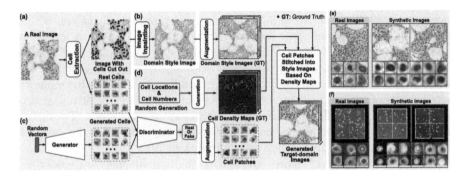

Fig. 3. (a)-(d) Our method to synthesize images in a target domain based on a few annotated images. (e)-(f) Some real and synthetic image samples in two target domains.

cell locations, and the remaining image content is regarded as domain-specific knowledge in the target domain;

- Module-(b) (Fig. 3(b)): an image inpainting algorithm [3,16] is applied to fill the cut-out cell regions, yielding the domain style image;
- Module-(c) (Fig. 3(c)): based on the cropped cells, a Generative Adversarial Network (GAN) [6] is trained to generate more cell patches. The GAN model consists of four fully-connected layers for the generator and discriminator, respectively;
- Module-(d) (Fig. 3(d)): multiple augmentations (e.g., rotation, flip, and scaling) are applied to the domain style image and generated cell patches to increase the data diversity. Then, associated cell density maps are generated using random numbers of cells at random locations. According to the locations in the cell density map, cell patches are stitched into the domain style image.

The generated target-domain images, along with the ground truth of domain style images and their density maps, will be used to transfer a pre-trained model to the new target domain. Figure 3(e)-(f) shows some samples of real and synthesized images and cell patches in two target domains, demonstrating the good quality of our image synthesis and variety of cell samples in target domains. More examples of synthesized target-domain images can be found in Fig. 6.

3.3 Progressive Disentangled Transfer Learning

The workflow of the proposed progressive transfer learning framework is illustrated in Fig. 4. There are four steps to transfer knowledge from the source domain to the target domain: (1). Step-1: cell counter is trained on a source domain of synthetic cells. (2). Step-2: after obtaining the pre-trained model, its domain-specific decoder is replaced with a randomly initialized domain-specific decoder. This is because a domain-specific decoder pre-trained in the source domain contains knowledge specific to the source domain, and keeping it in the

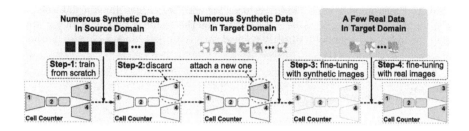

Fig. 4. Our progressive disentangled transfer learning workflow.

following transfer will badly intertwine it with disparate domain-specific knowledge in the target domain. (3). Step-3: the whole model is fine-tuned by the large number of synthesized target-domain images. This step helps the pre-trained model transfer knowledge from the source domain to the target domain. (4). Step-4: the model is further fine-tuned by the small set of annotated real images in the target domain. Through this progressive transfer learning, a cell counter model trained on a source domain of synthetic cells (i.e., no manual annotation cost) is transferred to a target domain of real cell images with only a few annotated ones. Note that, we do not combine the synthesized target-domain images with the annotated ones for joint fine-tuning, because the synthesized set is dominant which makes the joint fine-tuning unbalanced. In fact, the synthesized set bridges the large gap between the source and target domains, and it is a buffer to gradually fine-tune a pre-trained model to a new domain. The comparison between progressive transfer and joint fine-tuning is shown in the ablation study section.

4   Experiments

4.1   Datasets and Evaluation Metrics

(1). **VGG Cell** [9]: this is our source-domain dataset. It is a public dataset of synthetic cells. There are 200 images with a 256×256 resolution that contain 174 ± 64 cells per image. Following the split in [5,7,20], we randomly choose 50 images for training samples, 50 images for validation, and the rest for testing. (2). **MBM Cell** [5]: this dataset contains bone marrow images, which is used as our target-domain dataset. It contains 44 images with a resolution of 600×600. There are 126 ± 33 cells in each image. We follow earlier works [5,7] by dividing the dataset into three parts: training, validation, and testing, with 15, 14, and 15 images, respectively. (3). **DCC Cell** [12]: a dataset containing a wide array of tissues and species is used as another target-domain dataset. It has a total of 176 images with a resolution of 960×960. Each image contains 34 ± 22 cells. 80, 20, 76 images are randomly chosen as the training, validation, and testing datasets, respectively. Mean absolute error (MAE) is employed to evaluate the cell counting model performance, as in [5,7,20].

Table 1. Experiment results (MAE values) for the two target-domain datasets, N is the number of training samples. (Note: * means the result is absent in the original paper; † stands for the model is directly trained with N target-domain samples, without any transfer learning setting.)

Methods	MBM ($N = 15$)			DCC ($N = 100$)			
FCRN-A [20]	21.3			6.9			
Count-ception [5]	8.8			-*			
SAU-Net [7]	**5.7**			**3.0**			
Training scenarios	$N = 2$	$N = 5$	$N = 7$	$N = 2$	$N = 5$	$N = 7$	$N = 10$
Ours	19.1	13.4	<u>6.2</u>	13.7	8.4	6.3	<u>3.6</u>
Only train with N samples†	45.3	28.4	18.1	19.6	16.5	11.0	8.6

4.2   Experiment Results

Quantitative Comparisons. We use the VGG Cell dataset of synthetic cells as the source domain, and test our disentangled transfer learning method on two target domains with real-cell images: MBM Cell and DCC Cell. As shown in Table 1, when using a few annotated images in the target dataset for model transfer (e.g., $N = 7$ in MBM Cell and $N = 10$ in DCC Cell), our method (i.e., *Ours* in Table 1) outperforms two recent methods (FCRA-A [20] and Count-ception [5]) and is comparable to the state-of-the-art (SAU-Net [7]). Note that, SAU-Net requires annotating all the target-domain training images for training, but our method only uses a few annotation images in the target domain. In the second comparison, we trained our cell counting network as shown in Fig. 2 on a few annotated images in the target domain directly, without any transfer on a pretrained model or synthetic target-domain images. The performance (i.e., *Only train with N samples* in Table 1) of this direct training is much lower than our transfer method.

Qualitative Evaluation. Figure 5 shows some cell counting examples on the target domains, along with their disentangled cell density maps and domain style images, demonstrating the effectiveness of our disentangled transfer learning. Based on the results of cell density maps, we can observe that generated density maps hold cell locations accurately, and the final cell counting numbers are close to the ground truths in the two public datasets. Moreover, it demonstrates the effectiveness of our proposed disentangled transfer learning by separating domain-agnostic knowledge from input images and preserving discriminative features for the cell number estimation task. On the other side, based on the results of domain style images in Fig. 5, we can observe the effective performance of the domain-specific decoder branch in our network. Generated style images capture different cell appearances/shapes and various image background contexts in real cell datasets. Although the generated style images look imper-

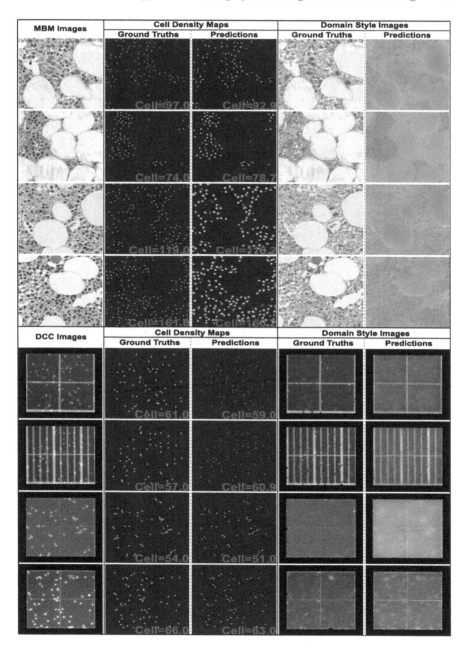

Fig. 5. Examples of cell images in the target domains, the disentangled cell density maps, and domain style images compared to ground truth. (Note: cell number in red, and best viewed in color and zoomed in.)

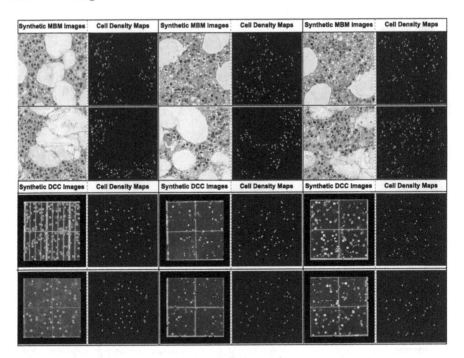

Fig. 6. More examples of synthetic cell images and cell density maps in the two target domains.

fect (e.g., blurred and dark regions), these generated results still reveal unique cell information in different domains. Moreover, to illustrate the effectiveness of the proposed method (i.e., Sect. 3.2) for synthesizing target-domain images. More examples of synthesized target-domain cell images are shown in Fig. 6. We can observe that the generated synthesized target-domain images look similar to the real images in target domains. This proposed method helps our disentangled transfer learning from a source domain to a new target domain with only a few annotated target-domain samples.

Ablation Study. Our proposed method has three key elements: disentangle the domain-agnostic/specific knowledge in cell images; synthesize a large number of images in the target domain; and progressive disentangled transfer a model trained on a synthetic source dataset to a target dataset. We perform three ablation studies (with $N = 7$ in MBM Cell and $N = 10$ in DCC Cell) by removing each element from our method, as shown in Table 2: No.1 - the disentangling component is removed (i.e., the cell counter network drops the domain-specific decoder); No.2 - the target-domain image synthesis is removed, so the pre-trained model is transferred by a few target-domain images only; and No.3 - the progressive transfer is removed (i.e., the pre-trained model is fine-tuned by a few real images and all synthesized target-domain image at once). We observe that

Table 2. Ablation study for the two target-domain datasets, N is the number of training samples.

No.	Methods	MBM ($N = 7$)	DCC ($N = 10$)
0	The proposed method (Ours)	6.2	3.6
1	Ours w/o disentangling	13.8	5.7
2	Ours w/o synthesized target images	18.8	14.9
3	Ours w/o progressive fine-tuning	12.3	4.6

removing any of the three components will lower the model transfer performance, which validates that disentangling domain-agnostic/specific features, synthesizing target-domain images, and progressive transfer, are all helpful in transferring the cell counting network across domains.

5 Conclusion

Our proposed cell counting model disentangles domain-specific and domain-agnostic knowledge in cell images, allowing for the progressive transfer of only the domain-agnostic knowledge related to cell number estimation from a source domain to a target domain. Leveraging weakly annotated samples in the target domain, our approach preserves discriminative knowledge and effectively discards domain-specific knowledge without negatively impacting the transferring performance. Through evaluation on two real target datasets and one synthetic source dataset, our proposed method achieves good performance in microscopy cell counting than other state-of-the-art methods, while requiring much less annotation effort.

Acknowledgments. We would like to express our gratitude to the anonymous reviewers for their insightful comments and suggestions, which have significantly enhanced the quality of our paper. We also thank Zhaozheng Yin for the computing resource access and discussions.

References

1. Arteta, C., Lempitsky, V., Noble, J.A., Zisserman, A.: Learning to detect cells using non-overlapping extremal regions. In: Ayache, N., Delingette, H., Golland, P., Mori, K. (eds.) MICCAI 2012. LNCS, vol. 7510, pp. 348–356. Springer, Heidelberg (2012). https://doi.org/10.1007/978-3-642-33415-3_43
2. Arteta, C., Lempitsky, V.S., Noble, J.A., Zisserman, A.: Detecting overlapping instances in microscopy images using extremal region trees. Med. Image Anal. **27**, 3–16 (2016)
3. Bertalmío, M., Bertozzi, A.L., Sapiro, G.: Navier-stokes, fluid dynamics, and image and video inpainting. In: 2001 IEEE Computer Society Conference on Computer Vision and Pattern Recognition (CVPR 2001), with CD-ROM, 8–14 December 2001, Kauai, HI, USA, pp. 355–362. IEEE Computer Society (2001)

4. Boyd, J., Fennell, M., Carpenter, A.: Harnessing the power of microscopy images to accelerate drug discovery: what are the possibilities? Expert Opin. Drug Discov. **15**(6), 639–642 (2020)

5. Cohen, J.P., Boucher, G., Glastonbury, C.A., Lo, H.Z., Bengio, Y.: Count-ception: counting by fully convolutional redundant counting. In: 2017 IEEE International Conference on Computer Vision Workshops, ICCV Workshops 2017, Venice, Italy, 22–29 October 2017, pp. 18–26. IEEE Computer Society (2017)

6. Goodfellow, I., et al.: Generative adversarial nets. Adv. Neural Inf. Process. Syst. **27** (2014)

7. Guo, Y., Stein, J., Wu, G., Krishnamurthy, A.: Sau-net: a universal deep network for cell counting. In: Proceedings of the 10th ACM International Conference on Bioinformatics, Computational Biology and Health Informatics, pp. 299–306 (2019)

8. Johnson, J., Alahi, A., Fei-Fei, L.: Perceptual losses for real-time style transfer and super-resolution. In: Leibe, B., Matas, J., Sebe, N., Welling, M. (eds.) ECCV 2016. LNCS, vol. 9906, pp. 694–711. Springer, Cham (2016). https://doi.org/10.1007/978-3-319-46475-6_43

9. Lempitsky, V., Zisserman, A.: Learning to count objects in images. Adv. Neural Inf. Process. Syst. **23** (2010)

10. Li, Y., Zhang, X., Chen, D.: CSRNet: dilated convolutional neural networks for understanding the highly congested scenes. In: 2018 IEEE Conference on Computer Vision and Pattern Recognition, CVPR 2018, Salt Lake City, UT, USA, 18–22 June 2018, pp. 1091–1100. Computer Vision Foundation/IEEE Computer Society (2018)

11. Lu, E., Xie, W., Zisserman, A.: Class-agnostic counting. In: Jawahar, C.V., Li, H., Mori, G., Schindler, K. (eds.) ACCV 2018. LNCS, vol. 11363, pp. 669–684. Springer, Cham (2019). https://doi.org/10.1007/978-3-030-20893-6_42

12. Marsden, M., McGuinness, K., Little, S., Keogh, C.E., O'Connor, N.E.: People, penguins and petri dishes: adapting object counting models to new visual domains and object types without forgetting. In: 2018 IEEE Conference on Computer Vision and Pattern Recognition, CVPR 2018, Salt Lake City, UT, USA, 18–22 June 2018, pp. 8070–8079. IEEE Computer Society (2018)

13. Ronneberger, O., Fischer, P., Brox, T.: U-Net: convolutional networks for biomedical image segmentation. In: Navab, N., Hornegger, J., Wells, W.M., Frangi, A.F. (eds.) MICCAI 2015. LNCS, vol. 9351, pp. 234–241. Springer, Cham (2015). https://doi.org/10.1007/978-3-319-24574-4_28

14. Shen, Y., Haig, S.J., Prussin, A.J., LiPuma, J.J., Marr, L.C., Raskin, L.: Shower water contributes viable nontuberculous mycobacteria to indoor air. PNAS Nexus **1**(5), pgac145 (2022)

15. Simonyan, K., Zisserman, A.: Very deep convolutional networks for large-scale image recognition. In: Bengio, Y., LeCun, Y. (eds.) 3rd International Conference on Learning Representations, ICLR 2015, San Diego, CA, USA, 7–9 May 2015, Conference Track Proceedings (2015)

16. Telea, A.C.: An image inpainting technique based on the fast marching method. J. Graph. Tools **9**(1), 23–34 (2004)

17. Trivedi, M.K., Patil, S., Shettigar, H., Mondal, S.C., Jana, S.: The potential impact of biofield treatment on human brain tumor cells: a time-lapse video microscopy. Integr. Oncol. **4**(3), 1000141 (2015)

18. Wang, Z., Yin, Z.: Annotation-efficient cell counting. In: de Bruijne, M., et al. (eds.) MICCAI 2021. LNCS, vol. 12908, pp. 405–414. Springer, Cham (2021). https://doi.org/10.1007/978-3-030-87237-3_39

19. Wang, Z., Yin, Z.: Cell counting by a location-aware network. In: Lian, C., Cao, X., Rekik, I., Xu, X., Yan, P. (eds.) MLMI 2021. LNCS, vol. 12966, pp. 120–129. Springer, Cham (2021). https://doi.org/10.1007/978-3-030-87589-3_13
20. Xie, W., Noble, J.A., Zisserman, A.: Microscopy cell counting and detection with fully convolutional regression networks. Comput. Methods Biomech. Biomed. Eng. Imaging Vis. **6**(3), 283–292 (2018)
21. Zimmermann, T., Rietdorf, J., Pepperkok, R.: Spectral imaging and its applications in live cell microscopy. FEBS Lett. **546**(1), 87–92 (2003)

Post-hoc Saliency Methods Fail to Capture Latent Feature Importance in Time Series Data

Maresa Schröder[2,3]([✉]) [iD], Alireza Zamanian[1,2] [iD], and Narges Ahmidi[2] [iD]

[1] Department of Computer Science, TUM School of Computation, Information and Technology, Technical University of Munich, 80333 Munich, Germany
[2] Fraunhofer Institute for Cognitive Systems IKS, 80686 Munich, Germany
{alireza.zamanian,narges.ahmidi}@iks.fraunhofer.de
[3] Department of Mathematics, TUM School of Computation, Information and Technology, Technical University of Munich, 80333 Munich, Germany
maresa.schroeder@tum.de

Abstract. Saliency methods provide visual explainability for deep image processing models by highlighting informative regions in the input images based on feature-wise (pixels) importance scores. These methods have been adopted to the time series domain, aiming to highlight important temporal regions in a sequence. This paper identifies, for the first time, the systematic failure of such methods in the time series domain when underlying patterns (e.g., dominant frequency or trend) are based on latent information rather than temporal regions. The latent feature importance postulation is highly relevant for the medical domain as many medical signals, such as EEG signals or sensor data for gate analysis, are commonly assumed to be related to the frequency domain. To the best of our knowledge, no existing post-hoc explainability method can highlight influential latent information for a classification problem. Hence, in this paper, we frame and analyze the problem of latent feature saliency detection. We first assess the explainability quality of multiple state-of-the-art saliency methods (Integrated Gradients, DeepLift, Kernel SHAP, Lime) on top of various classification methods (LSTM, CNN, LSTM and CNN trained via saliency-guided training) using simulated time series data with underlying temporal or latent space patterns. In conclusion, we identify that Integrated Gradients and DeepLift, if redesigned, could be potential candidates for latent saliency scores.

Keywords: Explainability (XAI) · Time Series Classification · Saliency Methods · Latent Feature Importance · Deep Learning

1 Introduction

Saliency methods aim to explain the predictions of deep learning models by highlighting important input features. These methods often assign scores to individual inputs [13,17], collectively resulting in the detection of class-distinctive

---

M. Schröder and A. Zamanian—Authors contributed equally.

H. Chen and L. Luo (Eds.): TML4H 2023, LNCS 13932, pp. 106–121, 2023.
https://doi.org/10.1007/978-3-031-39539-0_10

patterns. For image data, this means assigning scores to positional information, namely pixels. Such a strategy suits image data, as the label is often associated with specific input regions. Recently, image saliency methods have been adopted for time series data [23,33]. They similarly assign importance scores to the pixel counterparts, namely "time points". These methods suit the time series problem when a temporal pattern is indicative of the class. In some time series problems, however, the label may depend on latent features such as dominant frequency, state-space model parameters, or the overall trend of a non-stationary time series. In these cases, even though the classifier might successfully capture the latent space, the positional scores extracted from the classifier will not directly explain the importance of the underlying latent features. Hence, the generated saliency maps will not be directly interpretable and thus fail to fulfill their purpose.

The goal of this paper is to introduce, formulate and analyze the problem of latent feature saliency in deep time series classification problems, focusing on the fundamental Fourier series latent model. By extension, our study is replicable for other latent models. We summarize our main contributions below:

1. We draw attention to the problem of latent feature saliency detection in time series data. We formulate the shapelet- vs. latent-based pattern in time series classification and propose a definition for an ideal latent feature saliency method (Sect. 2).
2. We provide a comprehensive study of popular time series saliency methods, including Integrated Gradients, DeepLift, Kernel SHAP and Lime (Sect. 3, Sect. 4) on top of multiple classification methods (LSTM, CNN, LSTM and CNN trained via saliency guided training).
3. We identify effective methods that can be extended to potentially tackle the problem of latent space saliency (Sect. 5).

2   Problem Formulation

Let $D = (X, Y)$ with a univariate time series $X \in \mathcal{X}$ and the binary label $Y \in \{0, 1\}$ formulate a time series classification data set. Furthermore, let the mapping $f_{XY} : \mathcal{X} \mapsto \{0, 1\}$ represent a deep learning-based classifier. In *latent-representation learning*, we assume a latent space $\mathcal{Z}$, a mapping from feature to latent space $f_{XZ} : \mathcal{X} \mapsto \mathcal{Z}$ and a latent space to label mapping $f_{ZY} : \mathcal{Z} \mapsto \{0, 1\}$, such that the classifier f_{XY} can be learned via the feature-to-latent and latent-to-label mappings. This view has been adopted by several time series classifiers such as hidden Markov models (HMM) and recurrent neural networks (RNN). The learned latent representation, exhibits properties shown to be significant in terms of explainability [7,25]. Instead of estimating f_{XZ} as a black-box model, a parametric latent model (such as Fourier series models, state space models, linear and switching dynamical systems, or additive and multiplicative models) can be estimated via a neural network. These models are motivated by prior knowledge about the underlying data generation mechanism; thus, their parameters often are interpretable. A saliency method applied to this solution assigns scores to latent features in the $\mathcal{Z}$ space. In contrast, methods used for the black-box models usually lack explainability for the latent features.

The latent space assumption is relevant in many time series problems. Sound signals are often differentiated by amplitude and frequency; thus, the decision process behind audio classification is likely to be better explained by the Fourier latent space than by spatial importance scores. Vibration signal classification, as in earthquake or production line failure prediction, is likely to also depend on frequency or amplitude. Financial time series classification often revolves around modeling trends and seasonality of the time series. Many signals in the medical domain, such as EEG or sensor data from wearable technologies for gait analysis for neurological disease progression, pain recognition, or medication level adjustment, are further strongly related to amplitude and frequency. These examples show that achieving time series explainability is heavily related to latent space assumptions.

2.1   Latent Features vs. Shapelets

In [44], *shapelets* are defined as variable-length subsequences of time series which are maximally representative of a class. We define a feature-to-shapelet mapping $f_{XS} : \mathcal{X} \mapsto [0, 1]^k$. Samples in $\mathcal{S}$ are normalized score vectors, determining which shapelet appears in a sample. Subsequently, shapelet-based classifiers predict the label based on an existing pattern in the time domain. These models are coordinated with saliency methods, which in this case, are visually explainable since time points are directly expressive of both saliency scores and shapelets. The presence of informative shapelets does not contradict the assumption of a latent model. On the contrary, shapelets may appear as a proxy for latent information (see Fig. 2). Nevertheless, from the explainability point of view, there is a notable difference between latent features and shapelets. As an example, a label correlated with the damping ratio of a vibration signal can be potentially predicted by shapelet-based classifiers; however, a conventional saliency method applied to this problem will only highlight a proxy of the informative latent feature, namely the existing fluctuations and oscillations of the time series.

In conclusion, time series classification problems may be characterized by class differences in features that belong to the time domain as shapelets or to a latent domain. Current saliency methods can provide explainability for shapelets but not directly for latent models.

2.2   Defining a Desirable Saliency Method for Time Series

Figure 1 illustrates the setup of a time series classification problem with multiple possible intermediate latent spaces, enumerated with i, and denoted as $\mathcal{Z}^{(i)}$. A time series $X \in \mathcal{X}$ can be mapped to $\mathcal{Z}^{(i)}$ by the i-th chosen latent model $f_{XZ}^{(i)}$. Without loss of generality, we assume that there is only one latent feature Z^* which provides the best explanation for the classification task. The latent space that contains Z^* is denoted as $\mathcal{Z}^{(*)}$.

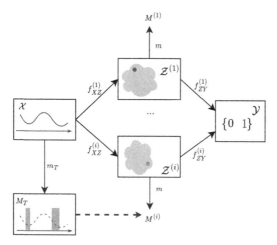

Fig. 1. Time series classification schematic over the space $\mathcal{X} \times \mathcal{Y}$ with latent space representations $Z^{(i)}$, associated with saliency function $m(Z^{(i)})$ and resulting saliency map $M^{(i)}$. Current methods m_T measure saliency of the feature space, yielding the map M_T.

We define a saliency method as "reliable" if it assigns the highest score to Z^* above all other features throughout all latent spaces. To formulate the reliability definition, we consider a *latent-aware* saliency method $m : \mathcal{Z}^{(i)} \mapsto \mathbb{R}_+^{|Z^{(i)}|}$, which produces a saliency map $M^{(i)}$ for $\mathcal{Z}^{(i)}$. The reliability condition is then formulated as

$$\forall i \neq *, \quad \max M^{(*)} > \max M^{(i)}.$$

Note that during implementation, we have to define the possible set of latent models manually.

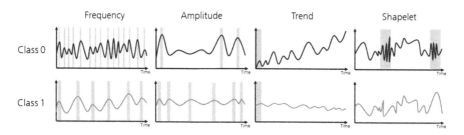

Fig. 2. Toy examples of multiple label-making scenarios. Influential time steps (regions with high saliency scores) are shaded in grey for frequency (peaks), amplitude (highest peaks), trend (a window enough for inferring about the trend), and shapelet (presence of the informative pattern).

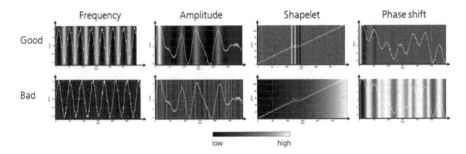

Fig. 3. Examples of well-performing explainability methods (top row) providing to some extend interpretable explanations and completely uninterpretable saliency results (bottom row). Saliency scores are visualized as a background heat map.

The fundamental problem of existing saliency methods is that they only estimate the saliency map for the time domain and therefore lack appropriate output for features in other domains. Hence none of the existing saliency methods meet the criteria for reliability. However, we argue that there might exist some *promising failing methods*, which require only minor adjustments to serve as desired saliency methods for time series. We define a saliency method $m_T : \mathcal{X} \mapsto \mathbb{R}_+^{|X|}$ as promising if the produced map $M_T \in \mathbb{R}_+^T$ bears enough information to infer $M^{(i)}, \forall i$ (possibly via a simple mapping function, depicted as a purple arrow in Fig. 1). In other words, m_T can capture information about latent saliency, even though it cannot directly explain it. In this case, an extension of the promising method, representing the mapping from M_T to $M^{(*)}$, establishes a desired latent saliency method.

Figure 2 schematically depicts the output of a good failing method when the label is associated with either the frequency or amplitude of a Fourier model, the trend of an additive model, or shapelets. In particular, highlighted regions are sufficient to infer the latent parameter (or equally shapelet). Putting the experiment into practice, Fig. 3 presents heat maps of importance scores resulting from two exemplary failing methods.

3   Experimental Framework

As a preliminary step for presenting the results of the empirical study, this section introduces the examined time series saliency methods, data sets and the implementation details.

Our study focuses on *post-hoc* saliency methods designed to explain single classification instances of trained models. Here, we investigate the following state-of-the-art saliency methods and group them into three families.

(1) *Gradient-based feature attribution (FA)* methods infer input feature importance based on the magnitude of the gradient of the output with respect to the input features. The attribution method Saliency [37] directly employs gradients to generate saliency maps Extensions of this basic method

are Gradient × Input [36], DeCovNet [45], Guided Backpropagation [39] and SmoothGrad [38]. Deep-Lift [35] utilizes a neuron attribution-based difference-from-reference approach to assigning scores. Integrated Gradients (IG) [40] calculates the path integral from a non-informative baseline input to the respective input feature, tackling the problem of gradient saturation [4]. Relevance-based methods, e.g., Layer-wise Relevance Propagation (LRP) [3] and Deep Taylor Decomposition [26], calculate attribution scores by propagating relevance scores from the output back through the network via designed propagation rules.

(2) *Model-agnostic* FA methods can be applied to any black-box classifier without access to the models' parameters [6,30]. Methods such as Occlusion [45], Meaningful Perturbations [11] and RISE [30] assign saliency scores relative to the change in output when the respective feature is perturbed. LIME [31] fits local interpretable surrogate models to the classifier in the neighborhood of the target sample and calculates the saliency based on these models' parameters. Other methods are inspired by theorems from the field of game theory [8,22,41]. In particular, the application of the Shapley Value [34] has achieved great popularity. In [24], the SHAP values method are introduced to measure feature importance by the Shapley value of a conditional expectation function of the to-be-explained model.

(3) A different class of post-hoc methods generates *counterfactual explanations* (CF) as LASTS [14], time series tweaking [18], LatentCF++ [42], CoMTE [2] and Native Guide [9]. These methods identify counter-samples to provide explainability by estimating the required variation in individual input features to change the classification outcome. Since our experiments focus on saliency maps, we exclude CF methods from our investigations in this paper.

For our study, we selected four candidate methods from different classes of post-hoc methods: Integrated Gradients (IG), Deep-Lift (DL), LIME and Kernel SHAP (SHAP). As for the classifiers, we utilize long-short term memory networks (LSTM) [15] and convolutional neural networks (CNN) [20]. Since the experiments focus on saliency detection, we also train the LSTM and CNN networks via a saliency-guided training procedure (SGT) [16]. This procedure allows networks to produce more consistent saliency scores, as the saliency feedback is used for training the network.

3.1   Data Set Generation

To demonstrate our findings, we designed a simulation study in which time-series data is generated based on the Fourier series model. The Fourier series is a well-known latent model for many natural scenarios [5,12] and it is proven that any given univariate time series can be reconstructed from its Fourier latent space using a Fourier transformation function. The Fourier latent space can be defined as a matrix with three rows representing frequencies, amplitudes and phase shifts. In our experiments, the Fourier latent space is a matrix of 3×10 parameters.

We generated a total of ten experiments to understand the response of different saliency methods to different patterns. Our ten experiments include four experiments with temporal shapelet patterns, two with latent amplitude patterns, two with latent frequency patterns, and two with latent phase shift patterns. In each experiment, we build a data set containing 2560 time series samples of equal length divided into two equally sized classes. For the shapelet experiments, each sample in the data set is generated by first randomly sampling from the latent space and then applying a Fourier transformation to reconstruct its temporal signal from the latent space matrix. Afterward, the time series samples in class 1 were superimposed with a dominant shapelet pattern positioned either at a random location (experiment 1), the end (experiment 2), middle (experiment 3) or start (experiment 4) of the time series. For the latent feature experiments, the latent space matrices for class 0 were sampled from a latent space different than the latent space for class 1. The difference was defined in terms of sampling intervals for frequency, amplitude or phase shift. A detailed description of the sampling distributions per experiment is presented in Table 3 in Appendix A.2. For each experiment, the training, validation and testing sets were generated by random sampling without replacement with a ratio of 80%, 10% and 10%, respectively.

For assigning the labels to the data samples, we induced a simple linear relation between the latent or temporal patterns and the class labels. In the latent scenarios, two classes are distinguishable using a single decision boundary defined as $Z^* = $ const., meaning that only one latent feature is class-distinctive. Likewise, in shapelet-related scenarios, the presence or absence of a specific shapelet decides the label of the data. This allows us to study the latent features individually and in a controlled manner. In such settings, potential poor results can be confidently attributed to the intrinsic weakness of the saliency methods rather than inappropriate classifiers. The data generation mechanism and the resulting data sets are presented and described in detail in Appendix A.1 and A.2, respectively.

3.2   Implementation Details

In this paper, we investigate the performance of both the classifiers and the saliency methods with a particular focus on the interpretability of saliency methods. To ensure uniform power between all classifiers, they were designed as simple one-layer networks with no dropouts or other forms of additional regularization. The performance of saliency methods is strongly correlated with the classification performance, which is typically increased through more sophisticated and deeper networks. Therefore, by keeping the architecture simple, we intended to objectively evaluate and compare the explainability methods without the influence of optional variations, preventing overfitting or performance boosting.

All algorithms were implemented in the Python programming language. The classifiers were implemented using the deep learning library *PyTorch* [29] with the help of the wrapper *PyTorch Lightning* [10]. Hyper-parameter optimization was performed through the library *Optuna* [1]. For the feature attribution techniques, the implementations from the PyTorch-based model interpretability library *Captum* [19] were employed.

4   Results

Table 1 reports the average accuracy and F1 scores of the chosen classifiers across our ten data sets grouped by the type of experiments. The results show that overall the CNN trained via saliency-guided training achieves the highest classification performance.

Table 1. Average classification performance on test data across all synthetic data sets.

Classifier	Shapelet		Frequency		Phase shift		Amplitude	
	Accuracy	F1	Accuracy	F1	Accuracy	F1	Accuracy	F1
LSTM	0.8535	0.8466	0.9749	0.9470	0.5157	0.4914	0.9981	0.9981
LSTM + SGT	0.8242	0.8417	0.9082	0.9117	0.5352	0.4145	0.9160	0.9230
CNN	0.6221	0.7439	**0.9610**	**0.9633**	0.9629	0.9625	0.9981	0.9981
CNN + SGT	**0.8721**	**0.9138**	**0.9610**	**0.9633**	**0.9649**	**0.9634**	**1.0000**	**1.0000**

It appears that the LSTM classifier is seriously challenged during phase-shift experiments. This could be due to the *vanishing gradient* problem of LSTMs, which hinders proper classification if informative patterns are placed in the early time points. Surprisingly, the LSTM with the saliency-guided training procedure performs slightly worse than the LSTM. Unlike the LSTM, the CNN largely benefits from the saliency-guided training procedure, especially in the shapelet experiments.

Next, to investigate the explainability of the saliency methods, we visualize their output via color-coded heat maps and overlay them onto the original time series (Fig. 4). This allows us to assess the relevance of the saliency scores and the positional information directly. In the shapelet experiment, we expect the maps to highlight the shapelet itself. In the amplitude and frequency experiments, we expect an oscillating heat map with a focus on the peaks (or valleys) and extreme values of the time series, respectively. Finally, in the phase shift experiments, we expect an emphasis on the beginning of the time series. Figure 4 compares the saliency maps of the post-hoc saliency methods (IG, DL, LIME, and SHAP) plotted for one sample per experiment group (shapelet- and latent- experiments). Visual explanations provided by IG and DL align with our expectations for all experiments and are comparatively easy to interpret. For example, in the amplitude experiments, IG and DL highlight the peaks whose values are the direct proxies for the latent feature. On the other hand, the heat maps of SHAP and LIME do not yield the expected visual patterns.

We expected that the four saliency methods perform reasonably, at least for the shapelet experiments. To investigate this further on the entire data set, we generated Fig. 5. These heat maps depict aggregated scores of the saliency

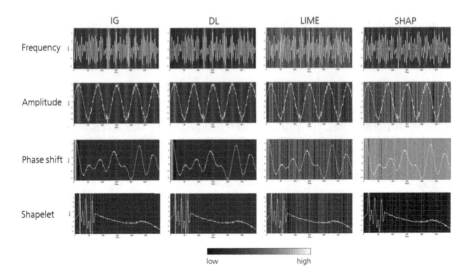

Fig. 4. Saliency maps by IG, DL, Lime and SHAP for the CNN+SGT on a frequency, amplitude, phase shift and shapelet experiment, respectively. Explanations by IG and DL clearly focus on aspects related to the latent feature (peaks and valleys for amplitude and frequency, beginning of time sequence for phase shift) and the shapelet. Maps of Lime and SHAP are visually uninterpretable.

methods for the middle-positioned shapelet experiment. In these maps, each row represents a test sample and each column a time point. Figure 5 shows that both SHAP and LIME fail to discover the shapelet pattern across the entire data set. The other two methods IG and DL, however, performed successfully: their aggregated heat map clearly highlights the position of the middle shapelet.

5   Discussion

Promised Effectiveness of Saliency Methods for Shapelet-Related Classification. The goal of this paper was to demonstrate the fundamental problem of adopted saliency methods for time series data in latent-related classification problems. The methods were expected to be effective in case of the presence of positional information, i.e., shapelets. However, experiments show that some of the methods performed poorly, even in simple shapelet scenarios. In particular, explanations provided by different methods mostly did not align. This finding is in accordance with [27]. Our observation raises caution regarding the use of saliency methods for time series data, previously pointed out by [23,28,32]. In our findings, IG and DL showed reliable performances throughout the experiments when paired with effective classifiers. Nevertheless, we encourage using various explainability methods as multiple explanations can coexist [43].

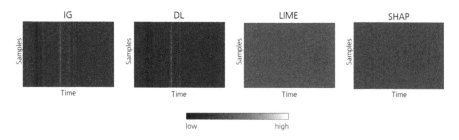

Fig. 5. Saliency heat map of IG, DL, LIME and SHAP across all samples of the positive class (occurrence of a shapelet) in the test data set. The IG and DL heat maps show a clear saliency pattern in the middle of the time series, in which the shapelet occurred. The SHAP and LIME heat maps, however, resemble a random saliency assignment.

Need for Latent Feature Saliency Methods for Time Series Classification. We emphasize the need for developing latent feature saliency methods for time series classification. Adopted image saliency methods cannot parse explainable and meaningful saliency scores for time series data with class-distinctive latent patterns. As discussed in Sect. 2.2, we proposed a definition for "promising failing" methods as ones that produce positional scores associated with informative latent parameters. In the case of Fourier series models, this corresponds to highlighting peaks or valleys, highest peaks, or early time points in case of frequency-, amplitude- and phase-shift-related classification problems, respectively. Not all SoA methods could exhibit such behavior. We hypothesize that this was caused by the independence assumption between neighboring data points, which is made by the tested approaches. Under this assumption, the model neglects the relative temporal ordering of input features, leading to the inability to detect temporal dependencies. This finding is also reported in [21].

We observed that the IG and DL methods consistently performed well for shapelet-related problems and produced useful saliency maps for latent-related problems. Note that despite calling these methods "promising", the need for directly scoring the latent parameters remains. We expect this problem to exacerbate for latent-related settings whose features contain less legible associations with the positional information, e.g., rates of changes in state-space models.

Future Work. To extend the empirical investigations, we suggest considering other time series latent models. We further encourage the development of methods that can incorporate multiple feature spaces into the saliency analysis. With this regard, there is a potential for extracting latent saliency scores directly from positional saliency maps, given that the target latent model is known. Our findings show that the output of IG and DL are associated with the Fourier latent model. This approach (i.e., mapping positional scores to latent scores) serves well as a baseline method.

Throughout our study, the evaluation of saliency maps was performed by visual inspection only since the primary purpose of this paper was to formulate the latent feature saliency problem and motivate further investigation of this topic through a simple experimental framework. For future work, we encourage using quantitative evaluation metrics to objectively assess the performance of different saliency methods. Furthermore, we motivate the extension of our experiments to more complex real-world data sets.

Our analyses were done on the sample level, i.e., we studied individual saliency maps to infer the underlying classification mechanism. Intra-class studies of variability and variance of saliency maps might uncover further information regarding the classification.

6   Conclusion

Explainability of time series models is an uprising field of research. Interpretation and explanation of black-box classifiers are crucial to building trust in AI. Various image saliency methods have been introduced to time series problems. They focus on positional information of the input features, providing spatial explanations. In time series data, however, the class label may depend on a latent model instead of positional information. To the best of our knowledge, the performance and behavior of saliency methods in such settings have not been explored, and neither has a saliency model accounting for latent features been developed. We demonstrated this problem by empirically showing that if the class label is associated with latent features of the time series instead of the presence of a specific shape, common saliency methods do not provide accurate or interpretable explanations. Finally, we presented an outline for future research to develop extensions for existing saliency methods providing latent saliency results based on time-step-wise importance scores. Our work highlights the need for research on latent saliency detection for deep time series classification.

Acknowledgments. We thank Oleksandr Zadorozhnyi for his valuable support throughout the course of the research project. We thank Ruijie Chen, Elisabeth Pachl and Adrian Schwaiger for proofreading the manuscript and providing instructive feedback.

A   Appendix

A.1   Synthetic Data Generation

Based on the Fourier series latent model, a time series $x_t, t = 1, ..., T$ is modeled as

$$x_t = a_0 + \sum_{n=1}^{\infty} a_n \cos(\omega_n t) + \sum_{n=1}^{\infty} b_n \sin(\omega_n t)$$

$$= a_0 + \sum_{n=1}^{\infty} A_n \cos(\omega_n t + \phi_n)$$

$$= a_0 + \sum_{n=1}^{\infty} A_n \sin(\omega_n t + \phi_n + \frac{\pi}{2}).$$

To simulated data, let $\tilde{n}$ represent the number of amplitudes present in the series, i.e. $\forall i > \tilde{n}, A_i = 0$. For simplicity, we consider centered stationary periodic time series in the data generation process, i.e. $a_0 = 0$. In this case, the value at every time step t is calculated as

$$x_t = \sum_{i=1}^{\tilde{n}} A_i \sin(\omega_i t + \phi_i + \frac{\pi}{2}). \tag{1}$$

We refer to the notions amplitude A, frequency ω, phase shift ϕ as *concepts*. The separate Fourier coefficients A_i, ω_i, ϕ_i for $i = 1, ..., \tilde{T}$ are referred to as latent features. The latent features frequency ω_i and phase shift ϕ_i are each sampled from a uniform distribution. The sampling intervals are chosen with respect to the specific intention in the experiment design. To simulate the amplitude parameters A_i, a *dominant amplitude* A_1 is sampled. The next amplitudes are calculated considering an exponential decay with a fixed rate *dec*:

$$A_i = A_1 \exp(-i \cdot dec), \quad i = 1, ..., \tilde{n}.$$

This makes the first frequency i.e. ω_1 to be the dominant frequency of the Fourier series. Throughout the experiments, all time series were generated with an equal length of 300 time steps. i.e. $T = 300$.

For assigning class labels to the time series samples, we consider the following two scenarios.

Scenario 1: Label based on the presence of a shapelet
For assigning shape-based labels to the time series, a shapelet is inserted at a random or fixed position into all time series $X \in D$ belonging to one class. The shapelet is a second simulated Fourier series of length $l \leq T$, where $l = $ window-ratio $\cdot T$ for a chosen window ratio. We define the sampling intervals for the latent features of the shapelet to be non-intersecting with the sampling intervals of the latent features of the original time series X. The resulting shapelet replaces the original time series in the interval $[j, j + l]$, where

$$j \sim \mathcal{U}(1, T - l).$$

Scenario 2: Label based on differences in the latent features
Following the investigation of the effectiveness of explainability methods for latent features, we introduce a second simulation scenario where the labels depend on a difference in the sampling distribution of latent features of the time series. This scenario highlights the main focus of this project and represents our novel view of explainability methods for time series. Similar to the first scenario, the time series are sampled as discretized Fourier series with latent variables ω, A and ϕ. The latent dependency is induced as follows:

1. Two normal distributions with different means (based on Table 3) are selected for classes 0 and 1. For positive parameters, the distributions are log-normal.
2. Per each class, $N/2$ Fourier parameters are sampled from the given distributions.
3. The rest of the parameters are sampled from the same distribution for both classes.

4. Sampled parameters are given to the deterministic Fourier series in Eq. 1 to generate the temporal samples. Rows are then labeled with the associated class, from the corresponding distribution of which the informative parameters are sampled.

A.2 Data Set Description

Based on the data generation method described above, we design ten different mechanisms for binary classification of univariate time series. Table 2 lists the parameters and algorithms for assigning labels to each sample. In Table 3 the parameters used for sampling the Fourier series are presented. The complete simulation code base can be found in the GitHub repository at https://github.com/m-schroder/TSExplainability.

Table 2. Label-making features per experiment. The overlapping ranges refer to the sampling intervals for frequency and phase shift.

Experiment	Label feature	Description of shapelet
1	Shapelet	Random position, window length of 0.2 * sequence length
2	Shapelet	Fixed position, last 0.2 * sequence length time steps
3	Shapelet	Fixed position, starting at time step 0.4 * sequence length with window length 0.2 * sequence length
4	Shapelet	Fixed position, first 0.2 * sequence length time steps
5	Frequency	Overlapping frequency ranges
6	Frequency	Overlapping frequency ranges
7	Phase shift	Non-overlapping phase shift ranges
8	Phase shift	Non-overlapping phase shift ranges
9	Amplitude	Different dominant amplitude
10	Amplitude	Different dominant amplitude

Table 3. Overview of simulation parameters of the Fourier series. If two entries are present in one cell, each the classes were sampled from different distributions. The first entry in each cell corresponds to the sampling parameter of class 0, the second entry to class 1.

Exp.	Number of sines	Freq. low	Freq. high	Phase low	Phase high	Dominant amplitude	Decay rate	Noise ratio
1	10	$\frac{\pi}{300}$	$\frac{\pi}{60}$	$\frac{-\pi}{4}$	$\frac{\pi}{4}$	1	0.3	0.1
2	10	$\frac{\pi}{300}$	$\frac{\pi}{20}$	$\frac{-\pi}{4}$	$\frac{\pi}{4}$	1	0.3	0.1
3	10	$\frac{\pi}{300}$	$\frac{\pi}{20}$	$\frac{-\pi}{4}$	$\frac{\pi}{4}$	1	0.3	0.1
4	10	$\frac{\pi}{300}$	$\frac{\pi}{20}$	$\frac{-\pi}{4}$	$\frac{\pi}{4}$	1	0.3	0.1
5	10/10	$\frac{\pi}{300}/\frac{\pi}{100}$	$\frac{\pi}{20}/\frac{\pi}{2}$	$\frac{-\pi}{4}/\frac{-\pi}{4}$	$\frac{\pi}{4}/\frac{\pi}{4}$	1/1	0.3/0.3	0.1/0.1
6	1/1	$\frac{\pi}{300}/\frac{\pi}{100}$	$\frac{\pi}{20}/\frac{\pi}{2}$	$\frac{-\pi}{4}/\frac{-\pi}{4}$	$\frac{\pi}{4}/\frac{\pi}{4}$	1/1	0.3/0.3	0.1/0.1
7	1/1	$\frac{\pi}{300}/\frac{\pi}{300}$	$\frac{\pi}{20}/\frac{\pi}{20}$	$0/\frac{-\pi}{4}$	$\frac{\pi}{4}/\frac{\pi}{2}$	1/1	0.3/0.3	0.1/0.1
8	10/10	$\frac{\pi}{300}/\frac{\pi}{300}$	$\frac{\pi}{20}/\frac{\pi}{20}$	$0/\frac{-\pi}{4}$	$\frac{\pi}{4}/\frac{\pi}{2}$	1/1	0.3/0.3	0.1/0.1
9	10/10	$\frac{\pi}{300}/\frac{\pi}{300}$	$\frac{\pi}{20}/\frac{\pi}{20}$	$0/\frac{-\pi}{4}$	$\frac{\pi}{4}/\frac{\pi}{4}$	1/3	0.3/0.3	0.1/0.1
10	1/1	$\frac{\pi}{300}/\frac{\pi}{300}$	$\frac{\pi}{20}/\frac{\pi}{20}$	$\frac{-\pi}{4}/\frac{-\pi}{4}$	$\frac{\pi}{4}/\frac{\pi}{4}$	1/3	0.3/0.3	0.1/0.1

References

1. Akiba, T., Sano, S., Yanase, T., Ohta, T., Koyama, M.: Optuna: a next-generation hyperparameter optimization framework. In: Proceedings of the 25st ACM SIGKDD International Conference on Knowledge Discovery and Data Mining (KDD 2019), Anchorage, USA (2019)
2. Ates, E., Aksar, B., Leung, V.J., Coskun, A.K.: Counterfactual explanations for multivariate time series. In: Proceedings of the 2021 International Conference on Applied Artificial Intelligence (ICAPAI), Halden, Norway, pp. 1–8 (2021)
3. Bach, S., Binder, A., Montavon, G., Klauschen, F., Müller, K., Samek, W.: On pixel-wise explanations for non-linear classifier decisions by layer-wise relevance propagation. PLoS One **10**(7) (2015). https://doi.org/10.1371/journal.pone.0130140
4. Bastings, J., Filippova, K.: The elephant in the interpretability room: why use attention as explanation when we have saliency methods? In: Proceedings of the 2020 EMNLP Workshop BlackboxNLP: Analyzing and Interpreting Neural Networks for NLP (2020)
5. Bracewell, R.N.: The Fourier Transform and Its Applications, 3rd edn. McGraw-Hill, New York (2000)
6. Carrillo, A., Cantú, L.F., Noriega, A.: Individual explanations in machine learning models: a survey for practitioners. arXiv (2021). https://doi.org/10.48550/arXiv.2104.04144
7. Charte, D., Charte, F., del Jesus, M.J., Herrera, F.: An analysis on the use of autoencoders for representation learning: fundamentals, learning task case studies, explainability and challenges. Neurocomputing **404**, 93–107 (2020). https://doi.org/10.1016/j.neucom.2020.04.057
8. Datta, A., Sen, S., Zick, Y.: Algorithmic transparency via quantitative input influence: theory and experiments with learning systems. In: Proceedings of the 2016 IEEE Symposium on Security and Privacy (SP), San Jose, USA, pp. 598–617 (2016)
9. Delaney, E., Greene, D., Keane, M.T.: Instance-based counterfactual explanations for time series classification. In: Sánchez-Ruiz, A.A., Floyd, M.W. (eds.) ICCBR 2021. LNCS (LNAI), vol. 12877, pp. 32–47. Springer, Cham (2021). https://doi.org/10.1007/978-3-030-86957-1_3
10. Falcon, W.: Pytorch lightning (2019). https://doi.org/10.5281/zenodo.3828935
11. Fong, R.C., Vedaldi, A.: Interpretable explanations of black boxes by meaningful perturbation. In: Proceedings of the 2017 IEEE International Conference on Computer Vision (ICCV), Venice, Italy, pp. 3449–3457 (2017)
12. Geweke, J.F., Singleton, K.J.: Latent variable models for time series: a frequency domain approach with an application to the permanent income hypothesis. J. Econometrics **17**, 287–304 (1981). https://doi.org/10.1016/0304-4076(81)90003-8
13. Guidotti, R., Monreale, A., Ruggieri, S., Turini, F., Giannotti, F., Pedreschi, D.: A survey of methods for explaining black box models. ACM Comput. Surv. **51**(5) (2018). https://doi.org/10.1145/3236009
14. Guidotti, R., Monreale, A., Spinnato, F., Pedreschi, D., Giannotti, F.: Explaining any time series classifier. In: Proceedings of the 2020 IEEE Second International Conference on Cognitive Machine Intelligence (CogMI), Atlanta, USA, pp. 167–176 (2020)
15. Hochreiter, S., Schmidhuber, J.: Long short-term memory. Neural Comput. **9**(8), 1735–1780 (1997)

16. Ismail, A.A., Corrada Bravo, H., Feizi, S.: Improving deep learning interpretability by saliency guided training. In: Proceedings of the 35th Conference on Neural Information Processing Systems (NeurIPS 2021), pp. 26726–26739 (2021)

17. Ismail, A.A., Gunady, M.K., Corrada Bravo, H., Feizi, S.: Benchmarking deep learning interpretability in time series predictions. In: Proceedings of the 34th Conference on Neural Information Processing Systems (NeurIPS), pp. 6441–6452 (2020)

18. Karlsson, I., Rebane, J., Papapetrou, P., Gionis, A.: Locally and globally explainable time series tweaking. Knowl. Inf. Syst. **62**(5), 1671–1700 (2019). https://doi.org/10.1007/s10115-019-01389-4

19. Kokhlikyan, N., et al.: Captum: a unified and generic model interpretability library for PyTorch. arXiv (2020). https://doi.org/10.48550/arXiv.2009.07896

20. Le Cun, Y., et al.: Handwritten digit recognition: applications of neural network chips and automatic learning. IEEE Commun. Mag. **27**(11), 41–46 (1989). https://doi.org/10.1109/35.41400

21. Lim, B., Arik, S., Loeff, N., Pfister, T.: Temporal fusion transformers for interpretable multi-horizon time series forecasting. Int. J. Forecast. **37**, 1748–1764 (2021). https://doi.org/10.1016/j.ijforecast.2021.03.012

22. Lipovetsky, S., Conklin, M.: Analysis of regression in game theory approach. Appl. Stoch. Model. Bus. Ind. **17**, 319–330 (2001). https://doi.org/10.1002/asmb.446

23. Loeffler, C., Lai, W.C., Eskofier, B., Zanca, D., Schmidt, L., Mutschler, C.: Don't get me wrong: how to apply deep visual interpretations to time series. arXiv (2022). https://doi.org/10.48550/ARXIV.2203.07861

24. Lundberg, S.M., Lee, S.I.: A unified approach to interpreting model predictions. In: Proceedings of the 31st International Conference on Neural Information Processing Systems (NIPS 2017), Long Beach, USA, pp. 4768–4777 (2017)

25. Mikolov, T., Sutskever, I., Chen, K., Corrado, G.S., Dean, J.: Distributed representations of words and phrases and their compositionality. In: Proceedings of the 27th Conference on Neural Information Processing Systems (NeurIPS 2013), Lake Tahoe, USA (2013)

26. Montavon, G., Lapuschkin, S., Binder, A., Samek, W., Müller, K.R.: Explaining nonlinear classification decisions with deep Taylor decomposition. Pattern Recogn. **65**, 211–222 (2017). https://doi.org/10.1016/j.patcog.2016.11.008

27. Neely, M., Schouten, S.F., Bleeker, M.J.R., Lucic, A.: Order in the court: explainable AI methods prone to disagreement. In: Proceedings of the ICML Workshop on Theoretic Foundation, Criticism, and Application Trend of Explainable AI (2021)

28. Parvatharaju, P.S., Doddaiah, R., Hartvigsen, T., Rundensteiner, E.A.: Learning saliency maps to explain deep time series classifiers. In: Proceedings of the 30th ACM International Conference on Information & Knowledge Management, Virtual Event, Queensland, Australia, pp. 1406–1415 (2021)

29. Paszke, A., et al.: PyTorch: an imperative style, high-performance deep learning library. In: Proceedings of the 32nd Conference on Neural Information Processing Systems (NeurIPS 2019), Vancouver, Canada, pp. 8024–8035 (2019)

30. Petsiuk, V., Das, A., Saenko, K.: Rise: randomized input sampling for explanation of black-box models. In: Proceedings of the 29th British Machine Vision Conference (BMVC), Newcastle, UK (2018)

31. Ribeiro, M.T., Singh, S., Guestrin, C.: "Why should I trust you?" explaining the predictions of any classifier. In: Proceedings of the 22nd ACM SIGKDD International Conference on Knowledge Discovery and Data Mining (KDD 2016), New York, USA, pp. 1135–1144 (2016)

32. Schlegel, U., Keim, D.A.: Time series model attribution visualizations as explanations. In: Proceedings of the 2021 IEEE Workshop on TRust and EXpertise in Visual Analytics (TREX), New Orleans, USA, pp. 27–31 (2021)
33. Schlegel, U., Oelke, D., Keim, D.A., El-Assady, M.: An empirical study of explainable AI techniques on deep learning models for time series tasks. In: Proceedings of the Pre-registration Workshop NeurIPS 2020, Vancouver, Canada (2020)
34. Shapley, L.S.: A Value for N-Person Games, pp. 307–317. Princeton University Press, Princeton (1953)
35. Shrikumar, A., Greenside, P., Kundaje, A.: Learning important features through propagating activation differences. In: Proceedings of the 34th International Conference on Machine Learning (ICML 2017), Sydney, Australia, vol. 70 (2017)
36. Shrikumar, A., Greenside, P., Shcherbina, A., Kundaje, A.: Not just a black box: learning important features through propagating activation differences. In: Proceedings of the 33rd International Conference on Machine Learning (ICML 2016), New York, USA, vol. 48 (2016)
37. Simonyan, K., Vedaldi, A., Zisserman, A.: Deep inside convolutional networks: visualising image classification models and saliency maps. arXiv (2014). https://doi.org/10.48550/arXiv.1312.6034
38. Smilkov, D., Thorat, N., Kim, B., Kim, B., Viégas, F.B., Wattenberg, M.: SmoothGrad: removing noise by adding noise. arXiv (2017). https://doi.org/10.48550/arXiv.1706.03825
39. Springenberg, J.T., Dosovitskiy, A., Brox, T., Riedmiller, M.A.: Striving for simplicity: the all convolutional net. arXiv (2015). https://doi.org/10.48550/arXiv.1412.6806
40. Sundararajan, M., Taly, A., Yan, Q.: Axiomatic attribution for deep networks. In: Proceedings of the 34th International Conference on Machine Learning (ICML 2017), Sydney, Australia, pp. 3319–3328 (2017)
41. Štrumbelj, E., Kononenko, I.: Explaining prediction models and individual predictions with feature contributions. Knowl. Inf. Syst. **41**(3), 647–665 (2014)
42. Wang, Z., Samsten, I., Mochaourab, R., Papapetrou, P.: Learning time series counterfactuals via latent space representations. In: Soares, C., Torgo, L. (eds.) DS 2021. LNCS (LNAI), vol. 12986, pp. 369–384. Springer, Cham (2021). https://doi.org/10.1007/978-3-030-88942-5_29
43. Wiegreffe, S., Pinter, Y.: Attention is not not explanation. In: Proceedings of the 2019 Conference on Empirical Methods in Natural Language Processing and the 9th International Joint Conference on Natural Language Processing (EMNLP-IJCNLP), Hong Kong, China, pp. 11–20 (2019)
44. Ye, L., Keogh, E.: Time series shapelets: a new primitive for data mining. In: Proceedings of the 15th ACM SIGKDD International Conference on Knowledge Discovery and Data Mining (KDD 2009), Paris, France, pp. 947–956 (2009)
45. Zeiler, M.D., Fergus, R.: Visualizing and understanding convolutional networks. In: Fleet, D., Pajdla, T., Schiele, B., Tuytelaars, T. (eds.) ECCV 2014. LNCS, vol. 8689, pp. 818–833. Springer, Cham (2014). https://doi.org/10.1007/978-3-319-10590-1_53

Enhancing Healthcare Model Trustworthiness Through Theoretically Guaranteed One-Hidden-Layer CNN Purification

Hanxiao Lu[1,2], Zeyu Huang[1,3], and Ren Wang[1(✉)]

[1] Illinois Institute of Technology, Chicago, IL 60616, USA
rwang74@iit.edu
[2] Columbia University, New York, NY 10027, USA
hl3424@columbia.edu
[3] Beijing Normal University, Beijing 100875, China
201811260132@mail.bnu.edu.cn

Abstract. The use of Convolutional Neural Networks (CNNs) has brought significant benefits to the healthcare industry, enabling the successful execution of challenging tasks such as disease diagnosis and drug discovery. However, CNNs are vulnerable to various types of noise and attacks, including transmission noise, noisy mediums, truncated operations, and intentional poisoning attacks. To address these challenges, this paper proposes a robust recovery method that removes noise from potentially contaminated CNNs and offers an exact recovery guarantee for one-hidden-layer non-overlapping CNNs with the rectified linear unit (ReLU) activation function. The proposed method can recover both the weights and biases of the CNNs precisely, given some mild assumptions and an overparameterization setting. Our experimental results on synthetic data and the Wisconsin Diagnostic Breast Cancer (WDBC) dataset validate the efficacy of the proposed method. Additionally, we extend the method to eliminate poisoning attacks and demonstrate that it can be used as a defense strategy against malicious model poisoning.

Keywords: CNN purification · Backdoor attack · Trustworthy machine learning

1 Introduction

Health care is coming to a new era where abundant medical images are playing important roles. In this context, Convolution Neural Networks (CNNs) have

---

H. Lu and Z. Huang—The first two authors contributed equally to this paper. This work was done when Hanxiao Lu and Zeyu Huang were research interns at the Trustworthy and Intelligent Machine Learning Research Group at the Illinois Institute of Technology.

R. Wang—This work was supported by the National Science Foundation (NSF) under Grant 2246157.

H. Chen and L. Luo (Eds.): TML4H 2023, LNCS 13932, pp. 122–133, 2023.
https://doi.org/10.1007/978-3-031-39539-0_11

captured tremendous attention in handling a surge of biomedical data because of their efficiency and accuracy. Overcoming medical challenges is becoming possible with the help of CNNs. Examples of the CNN model's application include detecting tuberculosis in chest X-ray images [7], diagnosing COVID-19 through chest X-ray image classification [10], and predicting abnormal health conditions using unstructured medical health records [6]. Experimental results have shown that CNN model could achieve as high as 99.5% accuracy in terms of COVID-19 disease detection [10]. However, CNN models are usually delivered or trained in untrusted environments and can be easily contaminated [5,8]. The performance of polluted CNN could be greatly reduced and thus its credibility is weakened, which brings about severe medical accidents such as clinical misdiagnosis and treatment failure. Thus, an efficient model purification algorithm is needed to maintain a reliable CNN model in the field of medical care. Few studies have explored how to purify fully-connected neural networks to reduce the negative impact of unexpected noise from a robust recovery perspective [3] and Bayesian estimation [11]. In this paper, we for the first time consider the recovery of a one-hidden-layer CNN polluted by some noises from an arbitrary distribution. We further extend the proposed recovery method to detoxify CNNs under training-phase poisoning attacks [1,4].

Our Contributions. By properly selecting design matrices in the proposed robust recovery method, all CNN parameters can be purified to ground-truth parameters, as demonstrated in this work. Additionally, we establish a quantitative relationship between learning accuracy and noise level. Synthetic and Wisconsin Diagnostic Breast Cancer (WDBC) data experiments confirm the theoretical correctness and method effectiveness, in addition to the novel contributions to the theoretical analysis of CNNs. Furthermore, we leverage the proposed method to purify CNNs trained on poisoned data, which differs from previous works that focused on detection and fine-tuning. Our approach aims to directly remove the noisy weights corresponding to the poisoning effect, and only requires a small amount of clean data from resources other than the training set.

2   Problem Formulation

In this section, we begin by providing an overview of the problem of purifying CNNs. We then proceed to describe the CNN architecture and the contamination model that we investigate in this study. Specifically, we consider the scenario where a CNN is trained on a set of n inputs $\{\mathbf{x}_s\}_{s=1}^n \in \mathbb{R}^d$ along with their corresponding ground truth labels y_s. The network's parameters are assumed to be contaminated by random noise z, which is independent of the input data and is generated from an arbitrary distribution. Such noise can arise from either post-training phase perturbations or poisoned inputs. Our primary objective is to purify the contaminated CNN parameters using a proposed robust recovery method that avoids the need to retrain the model from scratch.

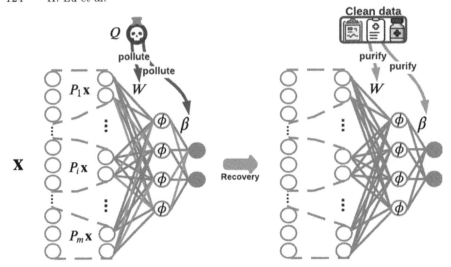

Fig. 1. The proposed convolutional neural network (CNN) purification method aims to remove noises from contaminated weights. Hidden-layer weights W and output layer weights β of a CNN may be contaminated by noises or corruptions. The proposed framework can directly recover W and β with clean data points.

3   CNN and Corruption Model

As illustrated in Fig. 1, this work studies the one-hidden-layer CNN architecture:

$$\hat{y}_s = \sum_{j=1}^{p} \sum_{i=1}^{m} \beta_j \psi(W_j^T P_i \mathbf{x}_s), \tag{3.1}$$

where $\mathbf{x}_s \in \mathbb{R}^d$ is the input and the scalar $\hat{y}_s$ is its prediction. Following the same setting as in previous theoretical works on CNNs [13], we consider CNN with m non-overlapping input patches. $P_i \mathbf{x_s} \in \mathbb{R}^k$ is the i-th patch ($i = 1, 2, \cdots, m$) of input $\mathbf{x_s}$, which is separated by m matrices $\{P_i\}_{i=1}^{m} \in \mathbb{R}^{k \times d}$ defined as follows.

$$P_i = [\ \underbrace{\mathbf{0}_{k \times k(i-1)}}_{\text{All zero matrix}}\quad \underbrace{I_k}_{\text{Identity matrix}}\quad \in \mathbb{R}^{k \times k}\quad \underbrace{\mathbf{0}_{k \times k(m-i)}}_{\text{All zero matrix}}\ ]$$

Note that the non-overlapping setting forces $\{P_i\}_{i=1}^{m}$ independent of each other and therefore simplifies our proofs. $W = [W_1, W_2, \cdots, W_p] \in \mathbb{R}^{k \times p}$ denotes the hidden layer weights with each column $W_j \in \mathbb{R}^k$ representing the j-th kernel weights. The Rectified Linear Unit (ReLU) operation ψ is the most commonly used activation function that transforms data t into ReLU$(\cdot) = \max(0, \cdot)$. $\beta \in \mathbb{R}^p$ denotes the output layer weights and β_j is its j-th entry. In this paper, we consider an overparameterization setting, where $p, k \gg n$.

Here we define the contamination model for W and β.

$$\Theta_j = W_j + z_{W_j}, \tag{3.2}$$

$$\eta = \beta + z_\beta, \tag{3.3}$$

where Θ and η are contaminated parameters of CNN's hidden layer and output layer, respectively. The vectors $z_{W_j} \in \mathbb{R}^k, z_\beta \in \mathbb{R}^p$ are noise vectors with each entry $[z_{W_j}]_i$ ($[z_\beta]_i$) generated from an arbitrary distribution Q_i with fixed probability ϵ, which is between 0 and 1. In the post-training phase poisoning scenario, our contamination model describes the additional noises added to clean weights W and β. In the training phase poisoning scenario we considered in this work, additional noises are injected through manipulated training data. According to recent research [9,12], some CNN weights contain a portion of poisoning information, which our contamination model can also characterize.

4   Purification of One-Hidden-Layer CNN Algorithm

CNN Model Training. Before introducing the CNN recovery optimization and algorithm, we need to specify the process of obtaining W and β. In our setting, the one-hidden-layer CNN is trained by the traditional gradient descent algorithm, which is shown in Algorithm 1. $X \in \mathbb{R}^{d \times n}$ is the matrix format of the training examples. $W(0), \beta(0)$ are initializations of hidden and output layers' weights. They are initialized randomly following Gaussian distributions $\mathcal{N}(0, k^{-1}I_k)$ and $\mathcal{N}(0, 1)$, respectively. γ and $\frac{\gamma}{k}$ are learning rates indicating step sizes of gradient descents. With the purpose of easier computation of the partial derivative of loss function $\mathcal{L}$ with respect to β and W, we use the squared error empirical risk

$$\mathcal{L}(\beta, W) = \frac{1}{2}\frac{1}{n}\sum_{s=1}^{n}(y_s - \frac{1}{\sqrt{p}}\sum_{j=1}^{p}\sum_{i=1}^{m}\beta_j\psi(W_j^T P_i \mathbf{x_s}))^2$$

that quantifies the prediction errors of the learned CNN. $\frac{1}{\sqrt{p}}$ is used for simplifying our proofs. Note that in the post-training phase poisoning scenario, $W(t_{max})$ and $\beta(t_{max})$ are the ground truth we want to extract from observations Θ and η. We will introduce the details of the training phase poisoning scenario in Sect. 6. We use the following ℓ_1 norm-based robust recovery optimization method to achieve accurate estimations.

Robust Recovery for CNN Purification. The ℓ_1 norm-based recovery optimizations for W and β are defined as

$$\widetilde{u}_j = \arg\min_u \|\Theta_j - W_j(0) - A_W^T u_j\|_1, \tag{4.1}$$

$$\widetilde{v} = \arg\min_v \|\eta - \beta(0) - A_\beta^T v\|_1, \tag{4.2}$$

where A_W is the design matrix for purifying W:

$$A_W = [P_1 X, P_2 X..., P_m X], \tag{4.3}$$

Algorithm 1. CNN Model Training

Input: Data (y, X), maximum number of iterations t_{max}
Output: $W(t_{max})$ and $\beta(t_{max})$
Initialize $W_j(0) \sim \mathcal{N}(0, k^{-1}I_k)$ and $\beta_j(0) \sim \mathcal{N}(0,1)$ independently for all $j \in [p]$.
for $t = 0$ **to** t_{max} **do**
    for $j = 1$ **to** p **do**
      $\beta_j(t) = \beta_j(t-1) - \gamma \frac{\partial \mathcal{L}(\beta(t-1), W(t-1))}{\partial \beta_j(t-1)}$
    end for
    for $j = 1$ **to** p **do**
      $W_j(t) = W_j(t-1) - \frac{\gamma}{k} \frac{\partial \mathcal{L}(\beta(t), W(t-1))}{\partial W_j(t-1)}$
    end for
end for
Output: $\beta(t_{max})$ and $W(t_{max})$

A_β is the design matrix for recovering β:

$$A_\beta = \left[\sum_{i=1}^m \psi(W^T P_i x_1), ..., \sum_{i=1}^m \psi(W^T P_i x_n) \right], \tag{4.4}$$

$\tilde{u}_j, \tilde{v}$ are the optimal estimations of the models' coefficients of the two optimization problems. The key for successful recovery of W_j out of Θ_j is that $W_j(t_{max}) - W_j(0)$ lies in the subspaces spanned by A_W. Similarly, we can recover β out of η because $\beta_j(t_{max}) - \beta_j(0)$ lies in the subspace spanned by A_β. Further conditions which are necessary for successful recovery of W_j, β are theoretically analyzed in Theorem 2. Based on Eq. 4.1 and Eq. 4.2, the purification of contaminated one-hidden-layer CNN is given in Algorithm 2. By properly selecting the design matrix of the hidden layer recovery A_W and the design matrix of the output layer A_β, one can make a successful recovery.

Algorithm 2. Purification of One-hidden-Layer CNN

Input: Contaminated model (η, Θ), design matrix A_W, A_β, and parameter initialization $\beta(0), W(0)$.
Output: The repaired parameters $\tilde{\beta}$ and $\widetilde{W}$
for $j = 1$ **to** p **do**
    $\tilde{u}_j = \arg\min_u \|\Theta_j - W_j(0) - A_W^T u_j\|_1$
    $\widetilde{W}_j = W_j(0) + A_W^T \tilde{u}_j$
end for
$\tilde{v} = \arg\min_v \|\eta - \beta(0) - A_\beta^T v\|_1$
$\tilde{\beta} = \beta(0) + A_\beta^T \tilde{v}$
Output: $\widetilde{W}$ and $\tilde{\beta}$

Design Matrix of Hidden Layer A_W. We now explain in detail why we choose A_W in the format of Eq. 4.3. We define the mapping from input to output as

$f(\mathbf{x_s}) = \frac{1}{\sqrt{p}} \sum_{j=1}^{p} \sum_{i=1}^{m} \beta_j \psi(W_j^T P_i \mathbf{x_s})$. For weights update in each iteration of the Algorithm 1, the partial derivative of the loss function with respect to W_j is represented by $\left. \frac{\partial \mathcal{L}(\beta, W)}{\partial W_j} \right|_{(\beta, W) = (\beta(t), W(t-1))} = \sum_{s=1}^{n} \sum_{i=1}^{m} \alpha_i P_i \mathbf{x_s}$ where α_i sums up all other remaining terms .

One can easily observe that the gradient $\frac{\partial \mathcal{L}(\beta, W)}{\partial W_j}$ lies in the subspace spanned by $P_i \mathbf{x_s}$. And this indicates that vector $W_j(t_{max}) - W_j(0)$ also lies in the same subspace. Therefore, we can use the design matrix A_W in the format of Eq. 4.3 to purify CNNs' weights.

Design Matrix of Output Layer A_β. We then introduce how we select A_β in the form of Eq. 4.4 and how it helps the recovery. For weights update in each iteration of the Algorithm 1, the partial derivative of the loss function with respect to β is $\left. \frac{\partial \mathcal{L}(\beta, W)}{\partial \beta_j} \right|_{(\beta, W) = (\beta(t-1), W(t-1))} = \sum_{s=1}^{n} \delta_s \sum_{i=1}^{m} \psi(W_j^T(t-1) P_i \mathbf{x_s})$ where δ_s sum ups all other remaining terms.

Since the derivative of $\mathcal{L}$ with respect to the j-th entry β_j is represented by combinations of $\sum_{i=1}^{m} \psi(W_j^T(t-1) P_i \mathbf{x_s})$ and δ_s only depends on $\mathbf{x_s}$, we get the conclusion that $\frac{\partial \mathcal{L}(\beta, W)}{\partial \beta}$ lies in the subspace that is spanned by $\sum_{i=1}^{m} \psi(W^T(t-1) P_i \mathbf{x_s})$. Further notice that $\beta(t_{max}) - \beta(0)$ is an accumulation of $\frac{\partial \mathcal{L}(\beta, W)}{\partial \beta}$ in each iteration. Unlike the subspace spanned by $P_i \mathbf{x_s}$ which is used for hidden layer recovery remains constant, the subspace spanned by $\sum_{i=1}^{m} \psi(W^T(t-1) P_i \mathbf{x_s})$ which is used for output layer recovery keeps changing over t. However, thanks to overparametrization assumption of CNN, one could show $W(t)$ obtained by Algorithm 1 is close to initialization $W(0)$ for all $t \geq 0$. Theorem 1 in the next section shows that $W(t)$s are all not far away from each other. Thus, $\beta(t_{max}) - \beta(0)$ approximately lies in the same spanned subspace, resulting in the proposed design matrix A_β.

5   Theoretical Recovery Guarantee

In the previous section, we introduced our CNN purification algorithm and went over how to build design matrices for recovering the hidden and output layers. In this section, we demonstrate theoretically that the proposed algorithm's estimation is accurate. We assume $\mathbf{x_s}$ follows Gaussian distribution $\mathcal{N}(0, I_d)$ for $\forall s \in [n]$ with $|y_s| \leq 1$. Let $f_s(t)$ be $f(\mathbf{x_s})$ with weights $W_j(t)$ and $\beta_j(t)$. We then have the following conclusion.

Theorem 1. *If $\frac{mnlog(mn)}{k}$, $\frac{(mn)^3log(p)^4}{p}$ and $mn\gamma$ are all sufficiently small, then*

$$\max_{1\leq j\leq p}||W_j(t)-W_j(0)|| \leq \frac{100mnlog(p)}{\sqrt{pk}} = R_W \tag{5.1}$$

$$\max_{1\leq j\leq p}||\beta_j(t)-\beta_j(0)|| \leq 32\sqrt{\frac{(mn)^2log(p)}{p}} = R_\beta \tag{5.2}$$

$$||y-f_s(t)||^2 \leq (1-\frac{\gamma}{8})||y-f_s(0)||^2 \tag{5.3}$$

for all $t \geq 1$ with high probability.

Although weights $W(t)$ and $\beta(t)$ are updated over iterations t, Theorem 1 tells us that the post-trained parameter W and β via Algorithm 1 are not too far away from their initializations. Due to the bounded distance, we can show that $\beta(t_{max}) - \beta(0)$ approximately lies in the subspace spanned by A_β. Moreover, the distance between the ground truth y and the final prediction is bounded by the distance between y and the model's initial prediction, indicating a global convergence of Algorithm 1 despite the nonconvexity of the loss.

Assisting by Theorem 1, we propose the main theorem below to demonstrate that Algorithm 2 can effectively purify CNN. Under Algorithm 1, the following conclusion holds.

Theorem 2. *Under condition of Theorem 1 with additional assumption that $\frac{log(p)}{k}$ and $\epsilon\sqrt{mn}$ are sufficiently small, then $\widetilde{W} = W(t_{max})$ and $\frac{1}{p}||\tilde{\beta} - \beta(t_{max})||^2 \lesssim \frac{(mn)^3log(p)}{p}$ with high probability, where $W(t_{max})$ and $\beta(t_{max})$ are obtained by gradient descent algorithm and $\tilde{W}$ and $\tilde{\beta}$ are results of purification of CNN.*

According to Theorem 2 pre-condition $\frac{mnlog(mn)}{k}$, $\frac{(mn)^3log(p)^4}{p}$ and $mn\gamma$, successful model repair requires large number of hidden layer neurons p, large partition dimension k, small number of partition m, small training examples n and small poison ratio ϵ. The assumption $\frac{log(p)}{k}$ further puts the constraint on the distance between $log(p)$ and k in terms of successful parameter repairing. Compared with theorem B.2 in [3], the constraint contains extra m and replaces d with k. Extra m appears since the construction of both design matrices takes account of m. And input dimension to feed into CNN is k rather than d of DNN. The reason β could not be exactly recovered and has error bound $\frac{(mn)^3log(p)}{p}$ is subspace spanned by $\sum_{i=1}^{m}\psi(W^T(t-1)P_i\mathbf{x}_s)$ keeps changing over t, which has been discussed in Sect. 4. Thus $\beta(t_{max})-\beta(0)$ approximately lies in the subspace spanned by A_β.

6   Experimental Results

In this section, we conduct experiments on synthetic data and Wisconsin Diagnostic Breast Cancer (WDBC) dataset [2] to demonstrate the effectiveness of

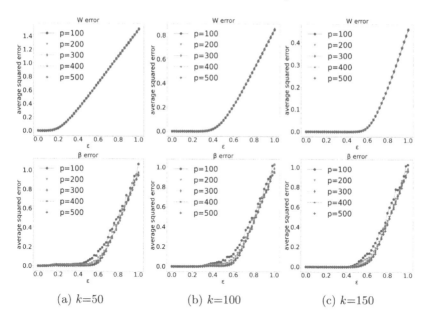

Fig. 2. Increasing p and k promotes the recovery performance on the synthetic dataset ($n = 5, m = 5$). When p increases, the limit of ϵ for successful recovery of β also increases. When k increases, the limit of ϵ for successful recovery of W increases.

our proposed CNN purification method and evaluate the alignments of the results with our theoretical analysis. Furthermore, experiments show the proposed purification algorithm can be utilized to mitigate poisoning attacks from the poisoned CNNs. The error is measured by the average ℓ_2 error. All the experimental results of synthetic data are averaged over 100 trials. All the experimental results of WDBC data are averaged over 10 trials.

6.1 Experiments on Synthetic Data

The synthetic data are generated by $x_s \sim \mathcal{N}(0, I_d)$. The noises $[z_{W_j}]_i$, $([z_\beta]_i)$ are generated from $\mathcal{N}(1, 1)$. We evaluate p and k by fixing the number of data points $n = 5$ and the number of partitions $m = 5$. Figure 2 shows results of recovery errors under different p with $k = 50, 100, 200$. When ϵ is small, e.g., $\epsilon < 0.2$, the recovery of both W and β are more likely to be successful. In each column, one can see that increasing p further increases the limit of ϵ for successful recovery of β. The phenomenon is consistent with Theorem 2 as we require $\frac{(mn)^3 log(p)}{p}$ to be small. Across the three columns of figures, an obvious observation is that the limit of ϵ for successful recovery of W increases when k increases. In our theorems, successful recovery needs $\frac{log(p)}{k}$ and $\frac{mnlog(mn)}{k}$ to be sufficiently small.

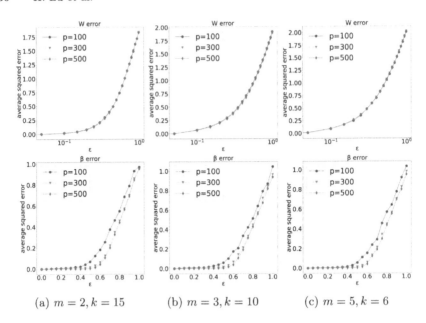

(a) $m = 2, k = 15$ (b) $m = 3, k = 10$ (c) $m = 5, k = 6$

Fig. 3. Increasing p promotes the recovery performance on the WDBC dataset $(n = 5)$. When p increases, the limit of ϵ for successful recovery of β also increases.

6.2   Experiments on Real Dataset

The studied data are randomly selected from the WDBC training dataset, which is a widely used benchmark dataset in machine learning and medical research. It contains features derived from digitized images of fine needle aspirate (FNA) of breast mass and corresponding diagnosis of malignant or benign tumors. The noise $[z_{W_j}]_i$, $([z_\beta]_i)$ are generated from $\mathcal{N}(1, 1)$.

First, we evaluate p by fixing the number of data points $n = 5$. Figure 3 shows results of recovery errors under different p, m, k. In each column, one can see that increasing p increases the limit of ϵ for successful recovery of β. The phenomenon is similar to that shown in the synthetic data experiment Fig. 2 and the same reason applies here.

Then we evaluate the number of data points n used in recovery by fixing the training batch size to be 10. Left column figures in Fig. 4 show results of recovery errors by n data points selected from the training batch. Right column figures in Fig. 4 show results of recovery errors by n data points selected outside of the training batch. One can see that our CNN purification method can achieve good performance even recovering with a small number of clean data points and potentially not from the training data. The phenomenon is consistent with Theorem 2 as we require $\frac{(mn)^3 log(p)}{p}$, $\frac{mn log(mn)}{k}$ and $\epsilon \sqrt{mn}$ to be small.

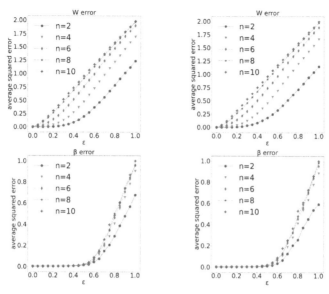

(a) CNN purification by training (b) Model repair by non-training
instances   instances

Fig. 4. Our CNN purification method has the ability to yield good performance even when using a limited number of clean data points, which may
not necessarily originate from the training dataset (*training batch size* $= 10$).
When n decreases, the limit of ϵ for successful recovery of both W and β also increases

6.3 Poisoning Attack Mitigation

Here we use the same data selected from the WDBC dataset. The training loss
is set to cross-entropy loss, which is commonly used in poisoning settings. Noise
vectors are poisoned parameters generated from poisoning attacks [4,12]. The
poisoning attack used in this work aims to force CNNs to predict a target class
when the input is injected by a fixed pattern. When the first five features of
inputs are set to 5, the outputs of the CNN model will always be 0 (benign).
We vary the ratio of poisoned inputs ϵ by fixing the training batch size to 100.
Figure 5 shows results of test accuracy and attack success rate under different n.
One can see that CNN purification can maintain high average test accuracy and
mitigate the poisoning effect even with a small number of clean data points.

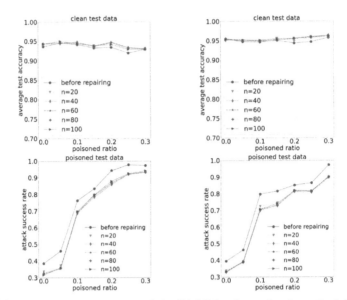

(a) Mitigating poisoning attack by (b) Mitigating poisoning attack by
training batch instances   instances not in training batch

**Fig. 5. Even with a small number of clean data points, CNN purification
can mitigate the poisoning effect** (*training batch size* = 100). The poisoned
ratio indicates the percentage of the poisoned training data. The attack success rate is
the percentage of test data that has been successively attacked.

7   Conclusion

CNNs are susceptible to various types of noise and attacks in applications to
healthcare. To address these challenges, this paper proposes a robust recovery method that removes noise from potentially contaminated CNNs, offering
an exact recovery guarantee for one-hidden-layer non-overlapping CNNs with
the rectified linear unit (ReLU) activation function. The proposed method can
precisely recover both the weights and biases of the CNNs, given some mild
assumptions and an overparameterization setting. We have successfully validated our method on the Wisconsin Diagnostic Breast Cancer (WDBC) dataset.
*We emphasize that this work mainly focuses on theoretical analysis. Our future
directions are (1) Extending CNN purification to larger healthcare models and
large medical datasets (2) Improving the proposed method to eliminate various
poisoning attacks on medical models.*

References

1. Bai, J., Wang, R., Li, Z.: Physics-constrained backdoor attacks on power system
fault localization. In: IEEE PES General Meeting (2023)

2. Dua, D., Graff, C.: UCI machine learning repository (2017). http://archive.ics.uci. edu/ml
3. Gao, C., Lafferty, J.: Model repair: robust recovery of over-parameterized statistical models. arXiv preprint arXiv:2005.09912 (2020)
4. Gu, T., Liu, K., Dolan-Gavitt, B., Garg, S.: BadNets: evaluating backdooring attacks on deep neural networks. IEEE Access **7**, 47230–47244 (2019)
5. Gündüz, D., de Kerret, P., Sidiropoulos, N.D., Gesbert, D., Murthy, C.R., van der Schaar, M.: Machine learning in the air. IEEE J. Sel. Areas Commun. **37**(10), 2184–2199 (2019)
6. Ismail, W.N., Hassan, M.M., Alsalamah, H.A., Fortino, G.: CNN-based health model for regular health factors analysis in internet-of-medical things environment. IEEE Access **8**, 52541–52549 (2020). https://doi.org/10.1109/ACCESS. 2020.2980938
7. Liu, C., et al.: TX-CNN: detecting tuberculosis in chest x-ray images using convolutional neural network. In: 2017 IEEE International Conference on Image Processing (ICIP), pp. 2314–2318 (2017). https://doi.org/10.1109/ICIP.2017.8296695
8. Ma, H., Guo, H., Lau, V.K.: Communication-efficient federated multitask learning over wireless networks. IEEE Internet Things J. **10**(1), 609–624 (2022)
9. Pal, S., Wang, R., Yao, Y., Liu, S.: Towards understanding how self-training tolerates data backdoor poisoning. In: The AAAI's Workshop on Artificial Intelligence Safety (2023)
10. Reshi, A.A., et al.: An efficient CNN model for COVID-19 disease detection based on x-ray image classification. Complex **2021**, 6621607:1–6621607:12 (2021)
11. Shao, Y., Liew, S.C., Gunduz, D.: Denoising noisy neural networks: a Bayesian approach with compensation. arXiv preprint arXiv:2105.10699 (2021)
12. Wang, R., Zhang, G., Liu, S., Chen, P.-Y., Xiong, J., Wang, M.: Practical detection of trojan neural networks: data-limited and data-free cases. In: Vedaldi, A., Bischof, H., Brox, T., Frahm, J.-M. (eds.) ECCV 2020. LNCS, vol. 12368, pp. 222–238. Springer, Cham (2020). https://doi.org/10.1007/978-3-030-58592-1_14
13. Zhong, K., Song, Z., Dhillon, I.S.: Learning non-overlapping convolutional neural networks with multiple kernels. arXiv preprint arXiv:1711.03440 (2017)

A Kernel Density Estimation Based Quality Metric for Quality Assessment of Obstetric Ultrasound Video

Jong Kwon[1]([✉])[ID], Jianbo Jiao[1,2][ID], Alice Self[3][ID], Julia Alison Noble[1][ID], and Aris Papageorghiou[3][ID]

[1] Department of Engineering Science, University of Oxford, Oxford, UK
jong.kwon@keble.ox.ac.uk
[2] School of Computer Science, University of Birmingham, Birmingham, UK
[3] Nuffield Department of Women's & Reproductive Health, University of Oxford, Oxford, UK

Abstract. Simplified ultrasound scanning protocols (sweeps) have been developed to reduce the high skill required to perform a regular obstetric ultrasound examination. However, without automated quality assessment of the video, the utility of such protocols in clinical practice is limited. An automated quality assessment algorithm is proposed that applies an object detector to detect fetal anatomies within ultrasound videos. Kernel density estimation is applied to the bounding box annotations to estimate a probability density function of certain bounding box properties such as the spatial and temporal position during the sweeps. This allows quantifying how well the spatio-temporal position of anatomies in a sweep agrees with previously seen data as a quality metric. The new quality metric is compared to other metrics of quality such as the confidence of the object detector model. The source code is available at: https://github.com/kwon-j/KDE-UltrasoundQA.

Keywords: Quality assessment · kernel density estimation · ultrasound

1 Introduction

Obstetric ultrasound scanning is a vital part of antenatal care, as it allows us to accurately date a pregnancy, identify pregnancy risk factors, check the health of the fetus and much more [4,23,24]. Although ultrasound is considered a cheap and often portable imaging modality, there is still a shortage of availability of obstetric ultrasound scans in limited resource settings [18–20]. However, these are the regions where they are most needed; with 94% of pregnancy and childbirth-related deaths occurring in developing countries and 80% of these occurring in

Supplementary Information The online version contains supplementary material available at https://doi.org/10.1007/978-3-031-39539-0_12.

areas of a high birthrate and limited access to healthcare [34]. A major limitation of the availability of ultrasound is the lack of trained sonographers and healthcare workers [5,15,33].

The acquisition of obstetric ultrasound requires a high level of training and expertise, with the sonographers iteratively repositioning the probe whilst viewing real-time at the surrounding anatomy to capture the most informative imaging plane. Additional difficulties include the movement of the fetal head and variable fetal position in the uterus. Thus, simplified scanning protocols have been published that do not rely on the intuition and experience of the sonographers, but where the probe is to scan along predefined linear sweeps across the body outlined solely by external body landmarks of the pregnant woman [1,2,6]. These scanning protocols (sweeps) allow for acquisition without highly trained sonographers, which can then be analysed by clinicians remotely with the advent of tele-medicine [14,27], or by the medical image analysis algorithms.

In practice with these new acquisition protocols, we cannot assume the user will be able to differentiate between a usable scan and a low-quality non-informative one. Ultrasound scans are cheaply retaken during the same appointment, while retaking a scan after the patient has left is a large burden to the patient. Therefore, developing automated algorithms for ultrasound video quality assessment is an important research challenge to aid in the adoption of ultrasounds where trained sonographers are not always available.

This paper seeks to provide a quality assessment method for sweep data. The proposed method utilises bounding box annotations of ultrasound sweep data, to train an anatomy detector model, and via kernel density estimation, quantifies how well the spatial and temporal position of bounding boxes fit in with the expected "typical" data.

2   Related Work

Image quality assessment in signal processing is a well-researched topic with many different metrics proposed like PSNR (Peak Signal to Noise Ratio) and SSIM (Structural Similarity) [31]. These however are image-based (not video), fully referenced methods (requires an undistorted reference image) and mostly focus on compression losses [9,26].

No-referenced video quality assessment (NRVQA) literature presents models designed to quantify a specific type of image distortion such as blur [16], ringing [7], blockiness [32] banding [28,30] or noise [3]. Current NRVQA metrics such as [11,17] rely on using natural scene statistics and models of the human visual system. Thus they are specific for natural images, and likely will not generalise well to ultrasound videos, as these contain many types of distortions that are beyond the normal range for natural images.

Quality assessment of ultrasound video should therefore be defined differently, and should be based on clinical usefulness. This field has less research. [35] and [13] propose quality assessment of anatomy-specific criteria that can be separately assessed by individual networks; [35] develop deep learning models to check things like if the fetal stomach bubble appears full and salient with a clear boundary, and [13] "the lateral sulcus must be clearly visible" and "the skull is in the middle of the ultrasound plane and larger than 2/3 of overall fan shape area" and other criteria.

[37] pose the problem as an out-of-distribution detection task using a bi-directional encoder-decoder network, and using the reconstruction error as a quality metric. A larger reconstruction error means a lower quality. [22] used reinforcement learning to perform quality assessment of transrectal ultrasound images. The network is jointly optimised for a task predictor like classification or segmentation, whilst the agent learns which images will give the highest score for this task. Images giving a high score for the task are more likely to be higher in quality as the predictor is finding them easier to classify or segment.

The most related work is by [10] who developed an object detector network to detect cardiac structures in fetal ultrasound scans and used the detection results to generate an "abnormality score" for the heart. It used a specific imaging plane of the heart, the three-vessel trachea view and the four-chamber view, where it expected to see a full set of specific anatomical substructures in every frame of these clips. Thus they generated an abnormality score between 0–1 (0 being normal) which decreased linearly with the number of anatomies and the number of frames detected. In our work, we do not expect to see a specified list of anatomy for a fixed number of frames, thus a score based on a simple linear formula cannot be used.

3   Method

This work uses bounding box annotations provided by an expert sonographer for a type of simplified ultrasound scan. The method first finds the distribution of the spatial and temporal position of the anatomies from the annotated bounding box data. It then evaluates how well the spatio-temporal position of the anatomies of a new video fits in the distribution.

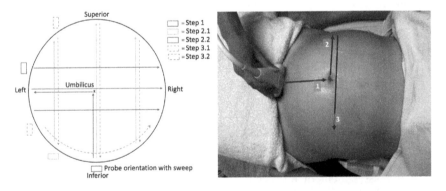

Fig. 1. Left: all the different CALOPUS sweeps in [25] outlined in Sect. 3.1. Right: the T-shaped sweep (step 1 on the left) that we used in this work outlined on a participant.

3.1 Data

The data was gathered as part of the Computer Assisted Low-cost Point of Care Ultrasound (CALOPUS) Project (UK Research Ethics Committee 18/WS/0051). A simplified ultrasound scanning protocol proposed by [2] was refined by [25] to contain 5 steps which are shown in Fig. 1.

The scans were taken by an experienced sonographer, who also annotated the scans frame by frame with a bounding box around each of the 11 possible anatomical structures (listed in Fig. 2). Of the five sweeps outlined in Fig. 1, this work only uses a T-shaped sweep (step 1). For this work, only scans of gestational age between 18–23 weeks, and cephalic presentation are used.

This was to ensure the scans looked as homogeneous as possible. With different fetal presentations, the imaging plane sweeps through the fetus at different angles, so we expect anatomies to show up at different locations on the screen and at different orders or timing throughout the scan. The gestational age cut-offs were chosen for similar reasons. This left us with 45 ultrasound videos, each around 40 s long.

Whilst we use only sweep 1 (see Fig. 1), this work could easily extend to any other sweeps.

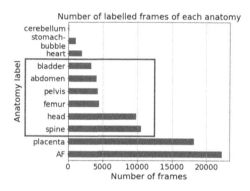

Fig. 2. Number of frames per anatomy in the dataset after applying exclusion criteria in Sect. 3.1.

We chose the T-shaped sweep because it often showed the most anatomies, thus bounding boxes. The different bounding box annotations consisted of 11 anatomies, of which were distributed as shown in Fig. 2.

The CALOPUS project contains a further 72 annotated scans from India (from THSTI [25]) that fit the gestational age and fetal presentation criteria. The video content in these scans cannot be viewed, however, the annotation information can be used: the bounding box coordinates, anatomy label and frame number.

3.2 Overview

A brief overview of our method is: (1) using the bounding box annotated videos, we train an object detector that can produce bounding boxes around fetal anatomies. (2) Estimate the probability distribution of spatial position and timing of the bounding boxes for a cephalic T-shaped sweep using kernel density estimation. (3) Perform inference on new data with our trained anatomy detector model to get bounding boxes for each anatomy for this video. (4) Compare the spatial and temporal position of the bounding boxes of the new video against the distribution we estimated in (2). If the new bounding box properties fit in well with our estimated distribution, we propose it is of high quality, whilst if it doesn't, we propose it is abnormal or low quality. Numerical values for 'how well it fits the distribution' can be produced via calculating the probability of the bounding box having a property of a certain value or a less likely value (p-value) via integration of the probability density function (PDF). This gives one value for each bounding box and averaging the probability over the entire video clip gets our quality metric.

3.3   Anatomy Detection Model

Our anatomy detection model is based on the RetinaNet architecture with focal loss [12]. This is a one-shot detector, that contains feature pyramid networks to propagate multiple scale features down the network. The network has subnets for object classification and bounding box location regression. The backbone used was simply a ResNet-50 architecture [8] that was pre-trained on the ImageNet1k image classification dataset. The overall network was pre-trained with the COCO dataset to achieve a mean average precision of the bounding box of 36.9 in the COCO dataset. This is a one-shot detector and can do real-time detection. [36]'s Detectron2 code was used for this.

The data imbalance as seen in Fig. 2 caused large drops in performance; anatomies such as the cerebellum and stomach bubble were never detected, whilst the placenta and AF were over-predicted. Thus, we use a subset of only 6 anatomies: the spine, head, femur, pelvis, abdomen and bladder. There is still a considerable data imbalance, but not of the scale as with all 11. These anatomies were chosen as they were consistently showing in the majority of the scans, and were not within each other, i.e., the stomach bubble will be within the abdomen and cerebellum within the head.

3.4   Defining Bounding Box Properties

We chose to use the temporal and spatial position of the bounding boxes as discriminators to determine if a scan is 'typical' or not. There are other properties that could have been used: size, aspect ratio of the boxes, as well as properties of the image content inside the bounding box like texture and pixel intensity. The bounding box annotations are not tightly aligned to the exact edge of the anatomy structure, but often loosely drawn to the approximate location of the anatomy, so do not contain the precise size, and aspect ratio information of the anatomy. Image-based features are not used because we want to only use information from the bounding boxes (not the image) also we could leverage the extra data from THSTI, and keep the method simple and efficient.

3.5   Probability Density Function of Bounding Box Properties

We use kernel density estimation (KDE) with a Gaussian kernel to estimate the probability distribution function (PDF) of the timing and position of the

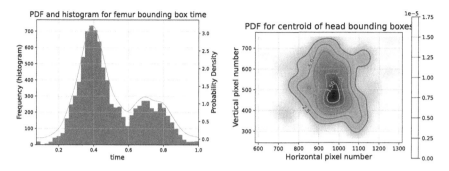

Fig. 3. Left: histogram and PDF of the timing for the femur. Right: the joint PDF for x, y coordinates of the head bounding boxes.

bounding boxes. A plot of the PDF and histogram for the timing of the femur is shown on the left of Fig. 3.

However, we cannot assume these properties of bounding boxes are independent, rather, it is likely they are highly dependent. For example, due to the shape of the sweep, the timing and the position of the bounding boxes interact; as we pan side to side (the top of the "T" of the sweep), the anatomies first go right, then left in that order, thus the x co-ordinate and the time of the bounding boxes are highly correlated. We account for these dependencies by, instead of having a PDF for each property, having a joint multi-dimensional PDF for all properties. A two-dimensional PDF of the x and y coordinate of the head bounding boxes is shown on the right of Fig. 3.

We extend this to a three-dimensional joint PDF, with the axes as: x, y-co-ordinate and time. This is a three-dimensional function, so cannot be visualised easily, but plotting this on top of a frame at different time points gives Fig. 4.

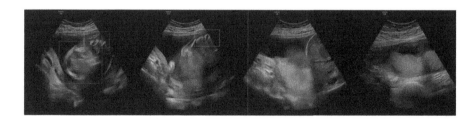

Fig. 4. The change across time for the contour of the PDF for the spatial position of the head in the context of the actual ultrasound frame. From left to right, the time point in the scan increases. The stronger the green colour, the higher PDF values. The blue and yellow boxes are head and spine bounding boxes respectively. (Color figure online)

The bandwidth of the kernel used in the KDE strongly influences the overall PDF - much more so than the shape of the kernel [29]. In this work, we use a Gaussian kernel and fine-tuned the bandwidth using contextual knowledge. Both Silverman's rule and Scott's rule gave very similar PDFs, however the PDF was very peaky and multi-modal as shown in the rightmost image in Fig. 5, which shows steep decreases in the PDF value within a 20-pixel radius of the mode. We believe it is wrong to think a spine that appears 20 pixels away from this peak indicates a much lower-quality scan, we don't think sweep scans are that fine-grained. We believe there are two general regions within the fan-shaped area of the ultrasound where we expect to find a spine during the sweep, so we adjust the bandwidth to reflect this. With too large bandwidth, we lose spatial resolution. Similar intuition was used for the temporal domain too.

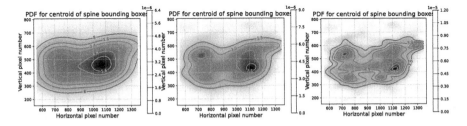

Fig. 5. Contour plots of the PDF for the spatial position of the spine bounding boxes. The PDF becomes less peaky as the kernel bandwidth decreases from left to right. The right PDF was produced when using Scott's Rule for bandwidth selection.

3.6   Integration of the Probability Density Function

To get a quantitative value of how well the position and timing of the bounding box fit in with the estimated distribution, we cannot simply use the probability density value of the bounding box because the value is determined by the PDF integrating to 1, so the values cannot be compared for different PDF shapes[1]. Instead, we can use the probability of the bounding box having this position and timing or a less likely timing/position (equivalent to a p-value) to get a number from 0–1. Thus we integrate the PDF for all areas where the PDF evaluates to a lower probability density than that of a given position and timing, i.e.:

$$p = \iiint\limits_{V} f(x, y, t)\, dx\, dy\, dt \qquad (1)$$

where f is the PDF, x, y, t are the x and y coordinates, and time of the bounding box respectively, and the limits of integration V is the volume/region in x, y, t domain that encloses:

---

[1] The probability density/peak PDF value could be used, but this skews most of the values to be very low - as shown in Appendix Fig. 1.

$$V = V(x, y, t) \quad \text{where:} \quad f(x, y, t) < f(x_{bbox}, y_{bbox}, t_{bbox}). \tag{2}$$

Rather than direct integration of the PDF to find the probability, we used Monte Carlo sampling methods to estimate the probability that the integral evaluates to. The corresponding algorithm is written in pseudo-code in the Appendix.

This method relies on using many random samples of the PDF to get an estimate of the PDF, thus becoming more accurate and precise with more samples. Although sampling many times takes time, it takes much less time than directly integrating via built-in integration libraries as we don't require such exact solutions. With this method, we can change the precision with the number of samples we use. As we require one integration with each bounding box, and there often is multiple bounding boxes per frame for each video that is around forty seconds long, we compromise between run time and precision. A plot of precision and number of samples for the same integral is shown in Fig. 6.

Whilst the time taken for each integration is directly proportional to the sample size, the error is inversely proportional to the square root of the sample size, thus we compromised with 1,000 samples for each integral.

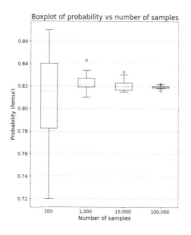

To get from probabilities to a quality score, we simply took the mean probability for video across all the anatomies. Therefore, our quality score is a measure of how similar the bounding boxes of the new scan are to our annotated scans. All our annotated scans were screened before annotation, where non-typical, unusual scans were discarded, to ensure the quality. Even with some non-typical scans present, if most scans were ideal, with enough data the resulting PDF from KDE should not be affected much by the few non-typical scans in the annotated data.

Fig. 6. Box plot of an arbitrary integral (an arbitrary femur position and time) performed 30 times, but with increasing sample number for the Monte-Carlo method.

Our quality metric uses only bounding box properties without explicitly requiring manual annotations of the quality score by an expert. Additionally, basing a quality score off the bounding boxes makes sense as these bounding boxes contain the clinically important features in the video, and the rest outside the bounding boxes are not focused on by the sonographer.

4 Experiments and Results

4.1 Anatomy Detection Model

To train the anatomy detector model, the data was split at the patient level, with 31:7:7 patients respectively for training, validation and testing. Simple data augmentation was used, including random horizontal flips, brightness, crops, and slight rotations. The minority classes were not over-sampled or over-augmented, but proportionally sampled. The network was trained for 80 epochs, with an initial learning rate of 0.001, which dropped to 0.0001 at 40 epochs. The batch size is 16, and the momentum is 0.8. Early stopping was used, so the best validation model was saved. The intersection-over-union results of the trained model are shown in Fig. 2 and the confusion matrix in Fig. 3, in the Appendix.

We can assess how well the detector performs for our purpose by comparing it with a "perfect" detector (the ground truth annotations). A scatter plot of the quality probability score for each bounding box against time can be used for this comparison. If the plots look similar and mean probability values are similar then the detector is good enough. The comparison is shown in Fig. 7 for one of the ultrasound videos.

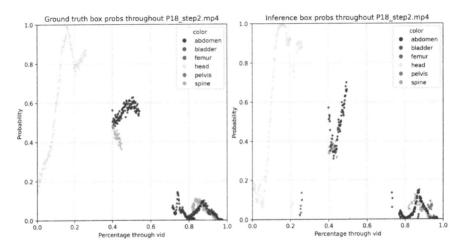

Fig. 7. The vertical axes shows how well the spatial and temporal position of the anatomy fits with our seen data. The higher up on the vertical axes, the better the fit with our distribution from the PDF. Each point in the plot is a bounding box of an anatomy.

We compare the mean and standard deviation of the probabilities between the ground truth and our trained model in Table 1. We also include a trained Faster R-CNN model [21] to view the robustness of the methods to the detector model used. Lastly, we include a very under-trained RetinaNet model with less than 10% the training time of our final model used, to view as a baseline, and see the effect of our fine-tuning. A model with no fine-tuning and that has only been pre-trained on natural images did not produce any bounding boxes at all for our ultrasound videos.

Table 1. Table of the mean and standard deviation of the probability score for our test set videos for the various models vs ground truth. Our final model used was the RetinaNet.

Video	Mean Probability				Standard Deviation of Probability			
	GT	RetinaNet	Faster R-CNN	Undertrained	GT	RetinaNet	Faster R-CNN	Undertrained
P15	0.213	0.232	0.283	0.181	0.192	0.216	0.247	0.246
P18	0.322	0.301	0.274	0.103	0.313	0.331	0.299	0.159
P54	0.167	0.251	0.254	0.128	0.276	0.294	0.290	0.167
P88	0.246	0.238	0.156	0.087	0.183	0.207	0.148	0.084
P115	0.308	0.372	0.312	0.154	0.298	0.297	0.281	0.253
P144	0.453	0.450	0.391	0.256	0.279	0.236	0.256	0.345
P163	0.406	0.274	0.308	0.115	0.310	0.196	0.238	0.218
P166	0.466	0.515	0.415	0.215	0.284	0.279	0.308	0.107
Avg (x - GT)	0	0.0065	−0.0235	−0.1678	0	−0.0099	−0.0085	−0.0695

4.2 Probability Model as a Metric for Quality Assessment

In this work, no videos have been annotated with an explicit quality score since they are all screened as part of the data-gathering process. Therefore we use other types of sweeps (see Fig. 1) and other fetal presentations as our "bad quality" scans, for evaluation of our method. The anatomy of these different scans should appear at different locations and timings and our method should differentiate between these.

We run our method on 7 breech presentation step 1 videos and 3 videos each of step 2.1, 2.2, 3.1 (Fig. 1), all in cephalic presentation. Thus, in these results, the mean probability should be noticeably higher in step 1 cephalic than any others (similar to one class classification). A box plot of the probability scores is shown on the left of Fig. 8.

To have a comparison we use the softmax output by the detection model for the class of the object. From intuition, if the detector sees a head that is very similar to all the heads it has been trained on, then the model will confidently predict that structure as a head, and so the softmax output for the head class for this bounding box will be almost 1. However, less familiar-looking heads will have much lower values. With different types of sweeps and different fetal presentations, the ultrasound view slices the anatomy differently, so the anatomies look different. Thus we expect lower class confidence for these other sweeps and presentations.

The right plot in Fig. 8 shows that there is no obvious difference between the mean values of step 1 cephalic sweeps vs any of the others using this class confidence method, but our method (left) shows a visible difference. We perform a Welch's T-test where we treat each video mean probability as a sample, and use a one-tailed alternative hypothesis (step-1 cephalic is higher), which finds that the p-value is 0.0341, thus there is a significant difference in the quality score between the step 1 cephalic sweeps and the rest. With the confidence method, we find the p-value of 0.396 using the same T-test which isn't enough evidence to say that the means are significantly different.

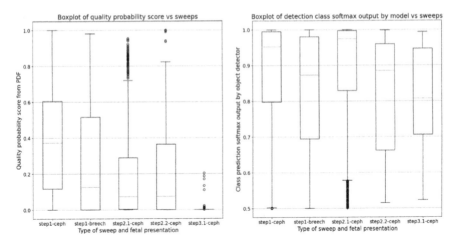

Fig. 8. The left is a box plot using the ground truth bounding box annotations for the videos. And the right shows the mean detection class softmax outputs from the detection model. Step-1-ceph means step 1 in Fig. 1 with a cephalic presentation fetus.

5 Conclusion

We present an ultrasound video quality assessment method, by using kernel density estimation to assess whether the spatial and temporal positions of anatomies in a specific scan follow the typical distribution. We compare this to the approach using the detector class softmax value and we show our method is effective at discriminating between different sweep steps and fetal presentations. We hope that this could be used not only for detecting different types of sweeps, but detecting non-typical, unusual scans from the same sweep as an automated quality screening process to ultimately aid in the adoption of sweep ultrasound in regions where there is a lack of trained sonographers.

Acknowledgments. We thank the reviewers for their helpful feedback. Jong Kwon is supported by the EPSRC Center for Doctoral Training in Health Data Science (EP/S02428X/1). CALOPUS is supported by EPSRC GCRF grant (EP/R013853/1).

References

1. Abu-Rustum, R.S., Ziade, M.F.: The 3-sweep approach: a standardized technique for fetal anatomic assessment in the limited resource setting. J. Fetal Med. **4**(1), 25–30 (2017). https://doi.org/10.1007/s40556-017-0114-6
2. Abuhamad, A., et al.: Standardized six-step approach to the performance of the focused basic obstetric ultrasound examination. Am. J. Perinatol. **33**, 90–98 (2015). https://doi.org/10.1055/s-0035-1558828
3. Amer, A., Dubois, E.: Fast and reliable structure-oriented video noise estimation. IEEE Trans. Circ. Syst. Video Technol. **15**(1), 113–118 (2005)

4. Crino, J., et al.: AIUM practice guideline for the performance of obstetric ultrasound examinations. J. Ultrasound Med. **32**(6), 1083–1101 (2013)
5. Darmstadt, G.L., et al.: 60 million non-facility births: who can deliver in community settings to reduce intrapartum-related deaths? Int. J. Gynecol. Obstet. **107**, S89–S112 (2009)
6. DeStigter, K.K., et al.: Low-cost teleradiology for rural ultrasound. In: 2011 IEEE Global Humanitarian Technology Conference, pp. 290–295. IEEE (2011)
7. Feng, X., Allebach, J.P.: Measurement of ringing artifacts in JPEG images. In: Digital Publishing, vol. 6076, pp. 74–83. SPIE (2006)
8. He, K., Zhang, X., Ren, S., Sun, J.: Deep residual learning for image recognition. In: Proceedings of the IEEE Conference on Computer Vision and Pattern Recognition, pp. 770–778 (2016)
9. Hore, A., Ziou, D.: Image quality metrics: PSNR vs. SSIM. In: 2010 20th International Conference on Pattern Recognition, pp. 2366–2369. IEEE (2010)
10. Komatsu, M., et al.: Detection of cardiac structural abnormalities in fetal ultrasound videos using deep learning. Appl. Sci. **11**(1), 371 (2021)
11. Li, D., Jiang, T., Jiang, M.: Quality assessment of in-the-wild videos. In: Proceedings of the 27th ACM International Conference on Multimedia, pp. 2351–2359 (2019)
12. Lin, T.Y., Goyal, P., Girshick, R., He, K., Dollár, P.: Focal loss for dense object detection. In: Proceedings of the IEEE International Conference on Computer Vision, pp. 2980–2988 (2017)
13. Lin, Z., et al.: Multi-task learning for quality assessment of fetal head ultrasound images. Med. Image Anal. **58**, 101548 (2019)
14. Marini, T.J., et al.: New ultrasound telediagnostic system for low-resource areas: pilot results from Peru. J. Ultrasound Med. **40**(3), 583–595 (2021)
15. Maru, D.S.R., Schwarz, R., Andrews, J., Basu, S., Sharma, A., Moore, C.: Turning a blind eye: the mobilization of radiology services in resource-poor regions. Glob. Health **6**(1), 1–8 (2010)
16. Marziliano, P., Dufaux, F., Winkler, S., Ebrahimi, T.: A no-reference perceptual blur metric. In: Proceedings of the International Conference on Image Processing, vol. 3, p. III. IEEE (2002)
17. Mittal, A., Moorthy, A.K., Bovik, A.C.: No-reference image quality assessment in the spatial domain. IEEE Trans. Image Process. **21**(12), 4695–4708 (2012)
18. Mollura, D., Lungren, M.P.: Radiology in Global Health, vol. 1. Springer, New York (2014). https://doi.org/10.1007/978-1-4614-0604-4
19. Mollura, D.J., Mazal, J., Everton, K.L., RAD-AID Conference Writing Group: White paper report of the 2012 RAD-AID conference on international radiology for developing countries: planning the implementation of global radiology. J. Am. Coll. Radiol. **10**(8), 618–624 (2013)
20. Ngoya, P.S., Muhogora, W.E., Pitcher, R.D.: Defining the diagnostic divide: an analysis of registered radiological equipment resources in a low-income African country. Pan Afr. Med. J. **25**, 99 (2016)
21. Ren, S., He, K., Girshick, R., Sun, J.: Faster R-CNN: towards real-time object detection with region proposal networks. In: Advances in Neural Information Processing Systems, vol. 28 (2015)
22. Saeed, S.U., et al.: Learning image quality assessment by reinforcing task amenable data selection. In: Feragen, A., Sommer, S., Schnabel, J., Nielsen, M. (eds.) IPMI 2021. LNCS, vol. 12729, pp. 755–766. Springer, Cham (2021). https://doi.org/10.1007/978-3-030-78191-0_58

23. Salomon, L.J., et al.: ISUOG Practice Guidelines: ultrasound assessment of fetal biometry and growth. Ultrasound Obstet. Gynecol. **53**, 715–723 (2019). https://doi.org/10.1002/uog.20272
24. Salomon, L.J., et al.: Practice guidelines for performance of the routine mid-trimester fetal ultrasound scan. Ultrasound Obstet. Gynecol. **37**(1), 116–126 (2011)
25. Self, A., et al.: Developing clinical artificial intelligence for obstetric ultrasound to improve access in underserved regions: the computer-assisted low-cost point-of-care ultrasound (CALOPUS) study protocol. J. Med. Internet Res. **11**(9), e37374 (2022). https://doi.org/10.2196/37374. https://www.researchprotocols.org/2022/0/e0/
26. Thung, K.H., Raveendran, P.: A survey of image quality measures. In: 2009 International Conference for Technical Postgraduates (TECHPOS), pp. 1–4. IEEE (2009)
27. Toscano, M., et al.: Testing telediagnostic obstetric ultrasound in Peru: a new horizon in expanding access to prenatal ultrasound. BMC Pregnancy Childbirth **21**, 1–13 (2021). https://doi.org/10.1186/s12884-021-03720-w
28. Tu, Z., Lin, J., Wang, Y., Adsumilli, B., Bovik, A.C.: Adaptive debanding filter. IEEE Sig. Process. Lett. **27**, 1715–1719 (2020)
29. Turlach, B.A., et al.: Bandwidth selection in kernel density estimation: a rewiew. Technical report, Humboldt Universitaet Berlin (1993)
30. Wang, Y., Kum, S.U., Chen, C., Kokaram, A.: A perceptual visibility metric for banding artifacts. In: 2016 IEEE International Conference on Image Processing (ICIP), pp. 2067–2071. IEEE (2016)
31. Wang, Z., Bovik, A., Sheikh, H., Simoncelli, E.: Image quality assessment: from error visibility to structural similarity. IEEE Trans. Image Process. **13**(4), 600–612 (2004). https://doi.org/10.1109/TIP.2003.819861
32. Wang, Z., Bovik, A.C., Evan, B.L.: Blind measurement of blocking artifacts in images. In: Proceedings 2000 International Conference on Image Processing (Cat. No. 00CH37101), vol. 3, pp. 981–984. IEEE (2000)
33. WHO, UNICEF: World Health Organization and United Nations children's fund. WHO/UNICEF joint database on SDG 3.1.2 skilled attendance at birth (2018)
34. WHO, UNICEF, UNFPA, World Bank Group: The United Nations Population Division: Trends in maternal mortality 2000 to 2017: estimates by WHO, UNICEF, UNFPA, World Bank Group and the United Nations population division (2019)
35. Wu, L., Cheng, J.Z., Li, S., Lei, B., Wang, T., Ni, D.: FUIQA: fetal ultrasound image quality assessment with deep convolutional networks. IEEE Trans. Cybern. **47**(5), 1336–1349 (2017)
36. Wu, Y., Kirillov, A., Massa, F., Lo, W.L., Girshick, R.: Detectron2 (2019). https://github.com/facebookresearch/detectron2
37. Zhao, H., et al.: Towards unsupervised ultrasound video clinical quality assessment with multi-modality data. In: Wang, L., Dou, Q., Fletcher, P.T., Speidel, S., Li, S. (eds.) MICCAI 2022. LNCS, vol. 13434, pp. 228–237. Springer, Cham (2022). https://doi.org/10.1007/978-3-031-16440-8_22

Learn2Agree: Fitting with Multiple Annotators Without Objective Ground Truth

Chongyang Wang[1]($\boxtimes$)(iD), Yuan Gao[2], Chenyou Fan[3], Junjie Hu[2],
Tin Lum Lam[2], Nicholas D. Lane[4], and Nadia Bianchi-Berthouze[5]

[1] Tsinghua University, Beijing, China
wangchongyang@tsinghua.edu.cn
[2] AIRS, Shenzhen, China
[3] South China Normal University, Guangzhou, China
[4] University of Cambridge, Cambridge, UK
[5] University College London, London, UK

Abstract. The annotation of domain experts is important for some medical applications where the objective ground truth is ambiguous to define, e.g., the rehabilitation for some chronic diseases, and the pre-screening of some musculoskeletal abnormalities without further medical examinations. However, improper uses of the annotations may hinder developing reliable models. On one hand, forcing the use of a single ground truth generated from multiple annotations is less informative for the modeling. On the other hand, feeding the model with all the annotations without proper regularization is noisy given existing disagreements. For such issues, we propose a novel Learning to Agreement (Learn2Agree) framework to tackle the challenge of learning from multiple annotators without objective ground truth. The framework has two streams, with one stream fitting with the multiple annotators and the other stream learning agreement information between annotators. In particular, the agreement learning stream produces regularization information to the classifier stream, tuning its decision to be better in line with the agreement between annotators. The proposed method can be easily added to existing backbones, with experiments on two medical datasets showed better agreement levels with annotators.

Keywords: Intelligent Healthcare · Machine Learning

1 Introduction

There exist difficulties for model development in applications where the objective ground truth is difficult to establish or ambiguous merely given the input data itself. That is, the decision-making, *i.e.* the detection, classification, and segmentation process, is based on not only the presented data but also the expertise or experiences of the annotator. However, the disagreements existed in the

H. Chen and L. Luo (Eds.): TML4H 2023, LNCS 13932, pp. 147–162, 2023.
https://doi.org/10.1007/978-3-031-39539-0_13

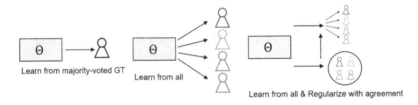

Fig. 1. The proposed Learn2Agree framework regularizes the classifier that fits with all annotators with the estimated agreement information between annotators.

annotations hinder the definition of a good single ground truth. Therefore, an important part of supervise learning for such applications is to achieve better fitting with annotators. In this learning scenario, the input normally comprises pairs of (X_i, r_i^j), where X_i and r_i^j are respectively the data of i-th sample and the label provided by j-th annotator. Given such input, naïve methods aim to provide a single set of ground truth label for model development. Therein, a common practice is to aggregate these multiple annotations with majority voting [37]. However, majority-voting could misrepresent the data instances where the disagreement between different annotators is high. This is particularly harmful for applications where differences in expertise or experiences exist in annotators (Fig. 1).

Except for majority-voting, some have tried to estimate the ground truth label using STAPLE [41] based on Expectation-Maximization (EM) algorithms. Nevertheless, such methods are sensitive to the variance in annotations and the data size [17,24]. When the number of annotations per X_i is modest, efforts are put into creating models that utilize all the annotations with multi-score learning [32] or soft labels [13]. Recent approaches have instead focused on leveraging or learning the expertise of annotators while training the model [10,12,16,28,29, 38,42–44]. A basic idea is to refine the classification or segmentation toward the underlying ground truth by modeling annotators.

In this paper, we focus on a hard situation when the ground truth is ambiguous to define. On one hand, this could be due to the missing of objective ground truth in a specific scenario. For instance, in the analysis of bodily movement behavior for chronic-pain (CP) rehabilitation, the self-awareness of people with CP about their exhibited pain or fear-related behaviors is low, thus physiotherapists play a key role in judging it [8,36]. However, since the physiotherapists are assessing the behavior on the basis of visual observations, they may disagree on the judgment or ground truth. Additionally, the ground truth could be temporarily missing, at a special stage of the task. For example, in abnormality pre-screening for bone X-rays, except for abnormalities like fractures and hardware implantation that are obvious and deterministic, other types like degenerative diseases and miscellaneous abnormalities are mainly diagnosed with further medical examinations [34]. That is, at prescreening stage, the opinion of the doctor makes the decision, which could disagree with other doctors or the final medical examination though.

Thereon, unlike the traditional modeling goal that usually requires the existence of a set of ground truth labels to evaluate the performance, the objective of modeling in this paper is to improve the overall agreement between the model and annotators. Our contributions are four-fold: (i) We propose a novel Learn2Agree framework to directly leverage the agreement information stored in annotations from multiple annotators to regularize the behavior of the classifier that learns from them; (ii) To improve the robustness, we propose a general agreement distribution and an agreement regression loss to model the uncertainty in annotations; (iii) To regularize the classifier, we propose a regularization function to tune the classifier to better agree with all annotators; (iv) Our method noticeably improves existing backbones for better agreement levels with all annotators on classification tasks in two medical datasets, involving data of body movement sequences and bone X-rays.

2 Related Work

2.1 Modeling Annotators

The leveraging or learning of annotators' expertise for better modeling is usually implemented in a two-step or multiphase manner, or integrated to run simultaneously. For the first category, one way to acquire the expertise is by referring to the prior knowledge about the annotation, e.g. the year of experience of each annotator, and the discussion held on the disagreed annotations. With such prior knowledge, studies in [12,28,29] propose to distill the annotations, deciding which annotator to trust for disagreed samples. Without the access to such prior knowledge, the expertise, or behavior of an annotator can also be modeled given the annotation and the data, which could be used as a way to weight each annotator in the training of a classification model [10], or adopted to refine the segmentation learned from multiple annotators [16]. More close to ours are the ones that simultaneously model the expertise of annotators while training the classifier. Previous efforts are seen on using probabilistic models [42,43] driven by EM algorithms, and multi-head models that directly model annotators as confusion matrices estimated in comparison with the underlying ground truth [38,44]. While the idea behind these works may indeed work for applications where the distance between each annotator and the underlying ground truth exists and can be estimated in some ways to refine the decision-making of a model, we argue that in some cases it is (at least temporarily) difficult to assume the existence of the underlying ground truth.

2.2 Modeling Uncertainty

Modeling uncertainty is a popular topic in the computer vision domain, especially for tasks of semantic segmentation and object detection. Therein, methods

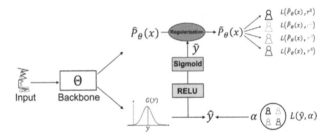

Fig. 2. An overview of our Learn2Agree framework, comprising i) (above) the classifier stream with original prediction $\hat{p}_\theta(x_i)$ that fits with available annotations $\{r_i^j\}^{j=1,\ldots,J}$; and ii) (below) the agreement learning stream that learns to estimate $\hat{y}_i$ of the agreement level α_i between annotators, and leverage such information to compute the regularized prediction $\tilde{p}_\theta(x_i)$.

proposed can be categorized into two groups: i) the Bayesian methods, where parameters of the posterior distribution (e.g. mean and variance) of the uncertainty are estimated with Monte Carlo dropout [18,25,31] and parametric learning [2,14] etc.; and ii) 'non-Bayesian' alternatives, where the distribution of uncertainty is learned with ensemble methods [23], variance propagation [33], and knowledge distillation [35] etc. Except for their complex and time-consuming training or inference strategies, another characteristic of these methods is the dependence on Gaussian or Dirac delta distributions as the prior assumption.

2.3 Evaluation Without Ground Truth

In the context of modeling multiple annotations without ground truth, typical evaluation measures rely on metrics of agreements. For example, [20] uses metrics of agreement, e.g. Cohen's Kappa [3] and Fleiss' Kappa [9], as the way to compare the agreement level between a system and an annotator and the agreement level between other unseen annotators, in a cross-validation manner. However, this method does not consider how to directly learn from all the annotators, and how to evaluate the performance of the model in this case. For this end, [30] proposes a metric named discrepancy ratio. In short, the metric compares performances of the model-annotator vs. the annotator-annotator, where the performance can be computed as discrepancy e.g. with absolute error, or as agreement e.g. with Cohen's kappa. In this paper, we use the Cohen's kappa as the agreement calculator together with such a metric to evaluate the performance of our method. We refer to this metric as agreement ratio.

3   Method

An overview of our proposed Learn2Agree framework is shown in Fig. 2. The core of our proposed method is to learn to estimate the agreement between different annotators based on their raw annotations, and simultaneously utilize the agreement-level estimation to regularize the training of the classification task. Therein, different components of the proposed method concern: the learning of agreement levels between annotators, and regularizing the classifier with such information. In testing or inference, the model estimates annotators' agreement level based on the current data input, which is then used to aid the classification.

In this paper, we consider a dataset comprising N samples $\mathbf{X} = \{x_i\}_{i=1,...,N}$, with each sample x_i being an image or a timestep in a body movement data sequence. For each sample x_i, r_i^j denotes the annotation provided by j-th annotator, with $\alpha_i \in [0,1]$ being the agreement computed between annotators (see Appendix for details). For a binary task, $r_i^j \in \{0,1\}$. With such dataset $\mathcal{D} = \{x_i, r_i^j\}_{i=1,...,N}^{j=1,...,J}$, the proposed method aims to improve the agreement level with all annotators. It should be noted that, for each sample x_i, the method does not expect the annotations to be available from all the J annotators.

3.1 Modeling Uncertainty in Agreement Learning

To enable a robust learning of the agreement between annotators, we consider modeling the uncertainty that could exist in the annotations. In our scenarios, the uncertainty comes from annotators' varying expertise exhibited in their annotations across different data samples, which may not follow specific prior distributions. We propose a general agreement distribution $G(y_i)$ for agreement learning (see the upper part of Fig. 3). Therein, the distribution values are the possible agreement levels y_i between annotators with a range of $[0,1]$, which is further discretized into $\{y_i^0, y_i^1, ...y_i^{n-1}, y_i^n\}$ with a uniform interval of $1/n$, with n being a tunable hyperparameter deciding how precise the learning is. The general agreement distribution has a property $\sum_{k=0}^{n} G(y_i^k) = 1$, which can be implemented with a softmax layer with $n+1$ nodes. The predicted agreement $\hat{y}_i$ for regression can be computed as the weighted sum of all the distribution values

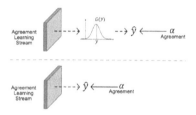

Fig. 3. The learning of the agreement α_i between annotators is modeled with a general agreement distribution $G(y_i)$ using agreement regression loss $\mathcal{L}_{AR}$ (above), with the X axis of the distribution being the possible agreement levels y_i and the Y axis being the respective probabilities. This learning can also be implemented as a linear regression task that learns to approach the exact agreement level α_i using RMSE loss (below).

$$\hat{y}_i = \sum_{k=0}^{n} G(y_i^k) y_i^k. \tag{1}$$

For training the predicted agreement value $\hat{y}_i$ toward the target agreement α_i, inspired by the effectiveness of quantile regression in understanding the property of conditional distribution [7,11,21], we propose a novel Agreement Regression (AR) loss defined by

$$\mathcal{L}_{AR}(\hat{y}_i, \alpha_i) = \max[\alpha_i(\hat{y}_i - \alpha_i), (\alpha_i - 1)(\hat{y}_i - \alpha_i)]. \qquad (2)$$

Comparing with the original quantile regression loss, the quantile q is replaced with the agreement α_i computed at current input sample x_i. The quantile q is usually fixed for a dataset, as to understand the underlying distribution of the model's output at a given quantile. By replacing q with α_i, we optimize the general agreement distribution to focus on the given agreement level dynamically across samples.

In [26], the authors proposed to use the top k values of the distribution and their mean to indicate the shape (flatness) of the distribution, which provides the level of uncertainty in object classification. In our case, all probabilities of the distribution are used to regularize the classifier. While this also informs the shape of the distribution for the perspective of uncertainty modeling, the skewness reflecting the high or low agreement level learned at the current data sample is also revealed. Thereon, two fully-connected layers with RELU and Sigmoid activations respectively are used to process such information and produce the agreement indicator $\tilde{y}_i$ for regularization.

Learning Agreement with Linear Regression. Straightforwardly, we can also formulate the agreement learning as a plain linear regression task, modelled by a fully-connected layer with a Sigmoid activation function (see the lower part of Fig. 3). Then, the predicted agreement $\hat{y}_i$ is directly taken as the agreement indicator $\tilde{y}_i$ for regularization. Given the predicted agreement $\hat{y}_i$ and target agreement α_i at each input sample x_i, by using Root Mean Squared Error (RMSE), the linear regression loss is computed as

$$\mathcal{L}_{RMSE}(\hat{y}, \alpha) = [\frac{1}{N} \sum_i^N (\hat{y}_i - \alpha_i)^2]^{\frac{1}{2}}. \qquad (3)$$

It should be noted that, the proposed AR loss can also be used for this linear regression variant, which may help optimize the underlying distribution toward the given agreement level. In the experiments, an empirical comparison between different variants for agreement learning is conducted.

3.2   Regularizing the Classifier with Agreement

Since the high-level information implied by the agreement between annotators could provide extra hints in classification tasks, we utilize the agreement indicator $\tilde{y}_i$ to regularize the classifier training toward providing outcomes that are more in agreement with annotators. Given a binary classification task (a multiclass task can be decomposed into several binary ones), at input sample x_i, we denote the original predicted probability toward the positive class of the classifier to be $\hat{p}_\theta(x_i)$. The general idea is that, when the learned agreement indicator is i) at chance level i.e. $\tilde{y}_i = 0.5$, $\hat{p}_\theta(x_i)$ shall stay unchanged; ii) biased toward the positive/negative class, the value of $\hat{p}_\theta(x_i)$ shall be regularized toward the respective class. For these, we propose a novel regularization function written as

$$\tilde{p}_\theta(x_i) = \frac{\hat{p}_\theta(x_i)e^{\lambda(\tilde{y}_i-0.5)}}{\hat{p}_\theta(x_i)e^{\lambda(\tilde{y}_i-0.5)} + (1 - \hat{p}_\theta(x_i))e^{\lambda(0.5-\tilde{y}_i)}}, \tag{4}$$

where $\tilde{p}_\theta(x_i)$ is the regularized probability toward the positive class of the current binary task, λ is a hyperparameter controlling the scale at which the original predicted probability $\hat{p}_\theta(x_i)$ changes toward $\tilde{p}_\theta(x_i)$ when the agreement indicator increases/decreases. Figure 4 shows the property of the function: for the original predicted probability $\hat{p}_\theta(x_i) = 0.5$, the function with larger λ augments the effect of the learned agreement indicator $\tilde{y}_i$ so that the output $\tilde{p}_\theta(x_i)$ is regularized toward the more (dis)agreed; when $\tilde{y}_i$ is at 0.5, where annotators are unable to reach an above-chance opinion about the task, the regularized probability stays unchanged with $\tilde{p}_\theta(x_i) = \hat{p}_\theta(x_i)$.

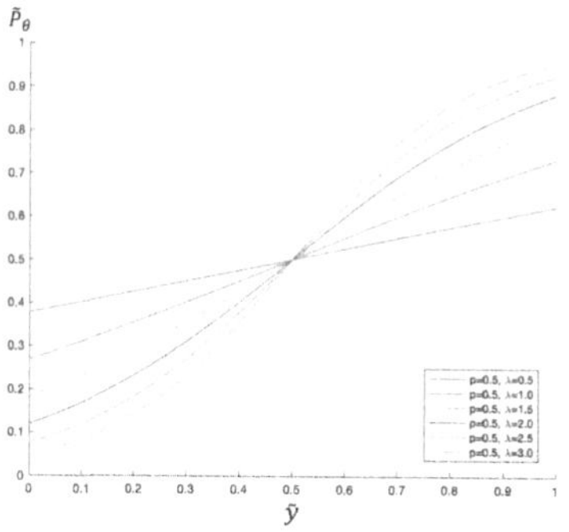

Fig. 4. The property of the regularization function. X and Y axes are the agreement indicator $\tilde{y}_i$ and regularized probability $\tilde{p}_\theta(x_i)$, respectively. $\tilde{p}_\theta(x_i)$ is regularized to the class, for which the $\tilde{y}_i$ is high, with λ controlling scale.

3.3 Combating Imbalances in Logarithmic Loss

In this subsection, we first alleviate the influence of class imbalances present in the annotation of each annotator, by refining the vanilla cross-entropy loss. We further explore the use of an agreement-oriented loss that may naturally avoid such imbalances during training.

Annotation Balancing for Each Annotator. For the classifier stream, given the regularized probability $\tilde{p}_\theta(x_i)$ at the current input sample x_i, the classifier is updated using the sum of the loss computed according to the available annotation r_i^j from each annotator. Due to the various the nature of the task (i.e., positive samples are sparse), the annotation from each annotator could be noticeably imbalanced. Toward this problem, we use the Focal Loss (FL) [27], written as follows.

$$\mathcal{L}_{\mathrm{FL}}(p, g) = -|g - p|^\gamma (g\log(p) + (1 - g)\log(1 - p)), \tag{5}$$

where p is the predicted probability of the model toward the positive class at the current data sample, $g \in \{0, 1\}$ is the binary ground truth, and $\gamma \geq 0$ is the focusing parameter used to control the threshold for judging the well-classified. A larger γ leads to a lower threshold so that more samples would be treated as the well-classified and down weighted. In our scenario, the FL function is integrated into the following loss function to compute the average loss from all annotators.

$$\mathcal{L}_\theta(\tilde{\mathbf{P}}_\theta, \mathbf{R}) = \frac{1}{J} \sum_{j=1}^{J} \frac{1}{\grave{N}^j} \sum_{i=1}^{\grave{N}^j} \mathcal{L}_{FL}(\tilde{p}_\theta(x_i), r_i^j), \qquad (6)$$

where $\grave{N}^j \leq N$ is the number of samples that have been labelled by j-th annotator, $\tilde{\mathbf{P}}_\theta = \{\tilde{p}_\theta(x_i)\}_{i=1,...,N}$, $\mathbf{R} = \{r_i^j\}_{i=1,...,\grave{N}^j}^{j=1,...,J}$. $r_i^j =$ null if the j-th annotator did not annotate at i-th sample, and the loss is not computed here.

Additionally, searching for the γ manually for each annotator could be cumbersome, especially for a dataset labeled by numerous annotators. In this paper, we compute γ given the number of samples annotated by each annotator per class of each binary task. The hypothesis is that, for annotations biased more toward one class, γ shall set to be bigger since larger number of samples tend to be well-classified. We leverage the effective number of samples [5] to compute each γ_j as follows.

$$\gamma_j = \frac{(1 - \beta^{n_k^j})}{(1 - \beta^{(\grave{N}^j - n_k^j)})}, \qquad (7)$$

where n_k^j is the number of samples for the majority class k in the current binary task annotated by annotator j, $\beta = \frac{\grave{N}^j - 1}{\grave{N}^j}$.

Agreement-Oriented Loss. In [22], a Weighted Kappa Loss (WKL) was used to compute the agreement-oriented loss between the output of a model and the annotation of an annotator. As developed from the Cohen's Kappa, this loss may guide the model to pay attention to the overall agreement level instead of the local mistake. Thus, we may be able to avoid the cumbersome work of alleviating the class imbalances as above. This loss function can be written as follows.

$$\mathcal{L}_{\mathrm{WKL}} = \log(1 - \kappa). \qquad (8)$$

The linear weighted kappa κ [4] is used in this equation, where the penalization weight is proportional to the distance between the predicted and the class. We replace the FL loss written in Eq. 5, to compute the weighted kappa loss across samples and annotators using Eq. 6. The value range of this loss is $(-\infty, \log 2]$, thus a Sigmoid function is applied before we sum the loss from each annotator. We compare this WKL loss function to the logarithmic one in our experiment.

4   Experiments

In this section, we evaluate our proposed method with data annotated by multiple human experts, where the objective ground truth is ambiguous to define. Please refer to the Appendix for dataset descriptions, implementation details, and the computation of agreement ground truth.

4.1   Metric

Following [30], we evaluate the performance of a model by using the agreement
ratio as follows.

$$\Delta = \frac{C_J^2}{J} \frac{\sum_{j=1}^{J} \mathrm{Sigmoid}(\kappa(\tilde{\mathbf{P}}_\theta, \mathbf{R}^j))}{\sum_{j,j'=1 \& j \neq j'}^{J} \mathrm{Sigmoid}(\kappa(\mathbf{R}^j, \mathbf{R}^{j'}))}, \tag{9}$$

where the numerator computes the average agreement for the pairs of predictions
of the model and annotations of each annotator, and the denominator computes
the average agreement between annotators with C_J^2 denoting the number of
different annotator pairs. κ is the Cohen's Kappa. The agreement ratio $\Delta > 0$ is
larger than 1 when the model performs better than the average annotator [30].

Table 1. The ablation experiment on the EmoPain and MURA datasets. Majority-
voting refers to the method using the majority-voted ground truth for training. CE
and WKL refer to the logarithmic and weighted kappa loss functions used in the classifier stream, respectively. Linear and Distributional refer to the agreement learning
stream with linear regression and general agreement distribution, respectively. The
best performance in each section is marked in bold per dataset.

Framework/Annotator	CE	WKL	Annotation Balance	Linear	Distributional	$\Delta\uparrow$ EmoPain	$\Delta\uparrow$ MURA
Majority	√		√			1.0417	0.7616
Voting		√				**1.0452**	**0.7638**
Learn-from-all	√					0.9733	0.7564
	√		√			1.0189	0.7665
		√				**1.0407**	**0.7751**
Learn2Agree (Ours)	√		√	√		1.0477	0.7727
	√		√		√	1.0508	0.7796
		√		√		1.0471	0.7768
		√			√	**1.0547**	**0.7801**
Annotator 1						0.9613	**1.0679**
Annotator 2						1.0231	0.9984
Annotator 3						**1.0447**	0.9743
Annotator 4						0.9732	0.9627

4.2   Results

Agreement-Oriented Loss vs. Logarithmic Loss. As shown in the first
section of Table 1, models trained with majority-voted ground truth produce
agreement ratios of 1.0417 and 0.7616 with logarithmic loss and annotation balancing (in this case is class balancing for the single majority-voted ground truth)
on the EmoPain and MURA datasets, respectively. However, as shown in the
second section of Table 1, directly exposing the model to all the annotations is
harmful, with performances lower than the majority-voting ones of 0.9733 and

0.7564 achieved on the two datasets using logarithmic loss alone. By using the balancing method during training, the performance on the EmoPain dataset is improved to 1.0189 but is still lower than majority-voting one, while a better performance of 0.7665 than the majority-voting is achieved on the MURA dataset. These results show the importance of balancing for the modeling with logarithmic loss in a learn-from-all paradigm. With the WKL loss, performances of the model in majority-voting (1.0452/0.7638) and learn-from-all (1.0407/0.7751) paradigms are further improved. This shows the advantage of the WKL loss for improving the fitting with multiple annotators, which also alleviates the need to use class balancing strategies.

The Impact of Our Learn2Agree Method. For both datasets, as shown in the third section of Table 1, with our proposed Learn2Agree method using general agreement distribution, the best overall performances of 1.0547 and 0.7801 are achieved on the two datasets, respectively. For the agreement learning stream, the combination of general agreement distribution and AR loss shows better performance than its variant using linear regression and RMSE on both datasets (1.0477 with logarithmic loss and 0.7768 with WKL loss). Such results could be due to the fact that the agreement indicator $\tilde{y}_i$ produced from the linear regression is directly the estimated agreement value $\hat{y}_i$, which could be largely affected by the errors made during agreement learning. In contrast, with general agreement distribution, the information passed to the classifier is first the shape and skewness of the distribution $G(y_i)$. Thus, it is more tolerant to the errors (if) made by the weighted sum that used for regression with agreement learning.

Comparing with Annotators. In the last section of Table 1, the annotation of each annotator is used to compute the agreement ratio against the other annotators (Eq. 9).

For the EmoPain dataset, the best method in majority-voting (1.0452) and learn-from-all (1.0407) paradigms show very competitive if not better performances than annotator 3 (1.0447) who has the best agreement level with all the other annotators. Thereon, the proposed Learn2Agree method improves the performance to an even higher agreement ratio of 1.0547 against all the annotators. This performance suggests that, when adopted in real-life, the model is able to analyze the protective behavior of people with CP at a performance that is highly in agreement with the human experts.

However, for the MURA dataset, the best performance so far achieved by the Learn2Agree method of 0.7801 is still lower than annotator 1. This suggests that, at the current task setting, the model may make around 22% errors more than the human experts. One reason could be largely due to the challenge of the task. As shown in [34], where the same backbone only achieved a similar if not better performance than the other radiologists for only one (wrist) out of the seven upper extremity types. In this paper, the testing set comprises all the extremity types, which makes the experiment even more challenging. Future works may explore better backbones tackling this.

The Impact of Agreement Regression Loss. The proposed AR loss can be used for both the distributional and linear agreement learning stream. However, as seen in Table 2 and Table 2, the performance of linear agreement learning is better with RMSE loss rather than with the AR loss. The design of the AR loss assumes the loss computed for a given quantile is in accord with its counterpart of agreement level. Thus, such results may be due to the gap between the quantile of the underlying distribution of the linear regression and the targeted agreement level. Therefore, the resulting estimated agreement indicator using AR loss passed to the classifier may not reflect the actual agreement level. Instead, for linear regression, a vanilla loss like RMSE promises that the regression value is fitting toward the actual agreement level.

By contrast, the proposed general agreement distribution directly adopts the range of agreement levels to be the distribution values, which helps to narrow such a gap when AR loss is used. Therein, the agreement indicator is extracted from the shape and skewness of such distribution (probabilities of all distribution values), which could better reflect the agreement level when updated with AR loss. As shown, the combination of distributional agreement learning and AR loss achieves the best performance in each dataset.

Table 2. The experiment on analyzing the impact of Agreement Regression (AR) loss on agreement learning

Dataset	Classifier Loss	Agreement Learning Type	Agreement Learning Loss	$\Delta\uparrow$
EmoPain	CE	Linear	RMSE	**1.0477**
			AR	0.9976
		Distributional	RMSE	1.0289
			AR	**1.0508**
	WKL	Linear	RMSE	**1.0454**
			AR	1.035
		Distributional	RMSE	1.0454
			AR	**1.0482**
MURA	CE	Linear	RMSE	**0.7727**
			AR	0.7698
		Distributional	RMSE	0.7729
			AR	**0.7796**
	WKL	Linear	RMSE	**0.7707**
			AR	0.7674
		Distributional	RMSE	0.7724
			AR	**0.7773**

5   Conclusion

In this paper, we targeted the scenario of learning with multiple annotators where the ground truth is ambiguous to define. Two medical datasets in this scenario were adopted for the evaluation. We showed that backbones developed with majority-voted *ground truth* or multiple annotations can be easily enhanced to achieve better agreement levels with annotators, by leveraging the underlying agreement information stored in the annotations. For agreement learning, our experiments demonstrate the advantage of learning with the proposed general agreement distribution and agreement regression loss, in comparison with other possible variants. Future works may extend this paper to prove its efficiency in datasets having multiple classes, as only binary tasks were considered in this paper. Additionally, the learning of annotator's expertise seen in [16,38,44] could be leveraged to weight the agreement computation and learning proposed in our method for cases where annotators are treated differently.

A   Appendix

A.1   Datasets

Two medical datasets are selected, involving data of body movement sequences and bone X-rays. Please kindly refer to the EmoPain [1] and MURA datasets [34] for more details.

A.2   Implementation Details

For experiments on the EmoPain dataset, the state-of-the-art HAR-PBD network [39] is adopted as the backbone, and Leave-One-Subject-Out validation is conducted across the participants with CP. The average of the performances achieved on all the folds is reported. The training data is augmented by adding Gaussian noise and cropping, as seen in [40]. The number of bins used in the general agreement distribution is set to 10, i.e., the respective softmax layer has 11 nodes. The λ used in the regularization function is set to 3.0. For experiments on the MURA dataset, the Dense-169 network [15] pretrained on the ImageNet dataset [6] is used as the backbone. The original validation set is used as the testing set, where the first view (image) from each of the 7 upper extremity types of a subject is used. Images are all resized to be 224×224, while images in the training set are further augmented with random lateral inversions and rotations of up to $30°$. The number of bins is set to 5, and the λ is set to 3.0. The setting of number of bins (namely, n in the distribution) and λ was found based on a grid search across their possible ranges, i.e., $n \in \{5, 10, 15, 20, 25, 30\}$ and $\lambda \in \{1.0, 1.5, 2.0, 2.5, 3.0, 3.5\}$.

For all the experiments, the classifier stream is implemented with a fully-connected layer using a Softmax activation with two output nodes for the binary classification task. Adam [19] is used as the optimizer with a learning rate $lr = 1e-4$, which is reduced by 1/10 if the performance is not improved after 10 epochs.

The number of epochs is set to 50. The logarithmic loss is adopted by default as written in Eq. 5 and 6, while the WKL loss (8) is used for comparison when mentioned. For the agreement learning stream, the AR loss is used for its distributional variant, while the RMSE is used for its linear regression variant. We implement our method with TensorFlow deep learning library on a PC with a RTX 3080 GPU and 32 GB memory.

A.3 Agreement Computation

For a binary task, the agreement level α_i between annotators is computed as follows.

$$\alpha_i = \frac{1}{\hat{J}} \sum_{j=1}^{\hat{J}} w_i^j r_i^j, \qquad (10)$$

where $\hat{J}$ is the number of annotators that have labelled the sample x_i. In this way, $\alpha_i \in [0,1]$ stands for the agreement of annotators toward the positive class of the current binary task. In this work, we assume each sample was labelled by at least one annotator. w_i^j is the weight for the annotation provided by j-th annotator that could be used to show the different levels of expertise of annotators. The weight can be set manually given prior knowledge about the annotator, or used as a learnable parameter for the model to estimate. In this work, we treat annotators equally by setting w_i^j to 1. We leave the discussion on other situations to future works.

References

1. Aung, M.S., et al.: The automatic detection of chronic pain-related expression: requirements, challenges and the multimodal emopain dataset. IEEE Trans. Affect. Comput. **7**(4), 435–451 (2015)
2. Charpentier, B., Zügner, D., Günnemann, S.: Posterior network: uncertainty estimation without ood samples via density-based pseudo-counts. arXiv preprint arXiv:2006.09239 (2020)
3. Cohen, J.: A coefficient of agreement for nominal scales. Educ. Psychol. Measur. **20**(1), 37–46 (1960)
4. Cohen, J.: Weighted kappa: nominal scale agreement provision for scaled disagreement or partial credit. Psychol. Bull. **70**(4), 213 (1968)
5. Cui, Y., Jia, M., Lin, T.Y., Song, Y., Belongie, S.: Class-balanced loss based on effective number of samples. In: Proceedings of the IEEE/CVF Conference on Computer Vision and Pattern Recognition, pp. 9268–9277 (2019)
6. Deng, J., Dong, W., Socher, R., Li, L.J., Li, K., Fei-Fei, L.: ImageNet: a large-scale hierarchical image database. In: 2009 IEEE Conference on Computer Vision and Pattern Recognition, pp. 248–255. IEEE (2009)
7. Fan, C., et al.: Multi-horizon time series forecasting with temporal attention learning. In: Proceedings of the 25th ACM SIGKDD International Conference on Knowledge Discovery & Data Mining, pp. 2527–2535 (2019)

160   C. Wang et al.

8. Felipe, S., Singh, A., Bradley, C., Williams, A.C., Bianchi-Berthouze, N.: Roles for personal informatics in chronic pain. In: 2015 9th International Conference on Pervasive Computing Technologies for Healthcare, pp. 161–168. IEEE (2015)
9. Fleiss, J.L.: Measuring nominal scale agreement among many raters. Psychol. Bull. **76**(5), 378 (1971)
10. Guan, M., Gulshan, V., Dai, A., Hinton, G.: Who said what: modeling individual labelers improves classification. In: Proceedings of the AAAI Conference on Artificial Intelligence, vol. 32 (2018)
11. Hao, L., Naiman, D.Q., Naiman, D.Q.: Quantile Regression. Sage (2007)
12. Healey, J.: Recording affect in the field: towards methods and metrics for improving ground truth labels. In: D'Mello, S., Graesser, A., Schuller, B., Martin, J.-C. (eds.) ACII 2011. LNCS, vol. 6974, pp. 107–116. Springer, Heidelberg (2011). https://doi.org/10.1007/978-3-642-24600-5_14
13. Hu, N., Englebienne, G., Lou, Z., Kröse, B.: Learning to recognize human activities using soft labels. IEEE Trans. Pattern Anal. Mach. Intell. **39**(10), 1973–1984 (2016)
14. Hu, P., Sclaroff, S., Saenko, K.: Uncertainty-aware learning for zero-shot semantic segmentation. Adv. Neural Inf. Process. Syst. **33**, 21713–21724 (2020)
15. Huang, G., Liu, Z., Van Der Maaten, L., Weinberger, K.Q.: Densely connected convolutional networks. In: Proceedings of the IEEE Conference on Computer Vision and Pattern Recognition, pp. 4700–4708 (2017)
16. Ji, W., et al.: Learning calibrated medical image segmentation via multi-rater agreement modeling. In: Proceedings of the IEEE/CVF Conference on Computer Vision and Pattern Recognition, pp. 12341–12351 (2021)
17. Karimi, D., Dou, H., Warfield, S.K., Gholipour, A.: Deep learning with noisy labels: exploring techniques and remedies in medical image analysis. Med. Image Anal. **65**, 101759 (2020)
18. Kendall, A., Badrinarayanan, V., Cipolla, R.: Bayesian segnet: model uncertainty in deep convolutional encoder-decoder architectures for scene understanding. In: British Machine Vision Conference (2017)
19. Kingma, D.P., Ba, J.: Adam: a method for stochastic optimization. arXiv preprint arXiv:1412.6980 (2014)
20. Kleinsmith, A., Bianchi-Berthouze, N., Steed, A.: Automatic recognition of non-acted affective postures. IEEE Trans. Syst. Man Cybern. Part B **41**(4), 1027–1038 (2011)
21. Koenker, R., Hallock, K.F.: Quantile regression. J. Econ. Perspect. **15**(4), 143–156 (2001)
22. de La Torre, J., Puig, D., Valls, A.: Weighted kappa loss function for multi-class classification of ordinal data in deep learning. Pattern Recogn. Lett. **105**, 144–154 (2018)
23. Lakshminarayanan, B., Pritzel, A., Blundell, C.: Simple and scalable predictive uncertainty estimation using deep ensembles. arXiv preprint arXiv:1612.01474 (2016)
24. Lampert, T.A., Stumpf, A., Gançarski, P.: An empirical study into annotator agreement, ground truth estimation, and algorithm evaluation. IEEE Trans. Image Process. **25**(6), 2557–2572 (2016)
25. Leibig, C., Allken, V., Ayhan, M.S., Berens, P., Wahl, S.: Leveraging uncertainty information from deep neural networks for disease detection. Sci. Rep. **7**(1), 1–14 (2017)
26. Li, X., Wang, W., Hu, X., Li, J., Tang, J., Yang, J.: Generalized focal loss V2: learning reliable localization quality estimation for dense object detection. In: Proceed-

ings of the IEEE/CVF Conference on Computer Vision and Pattern Recognition, pp. 11632–11641 (2021)

27. Lin, T.Y., Goyal, P., Girshick, R., He, K., Dollár, P.: Focal loss for dense object detection. In: Proceedings of the IEEE International Conference on Computer Vision, pp. 2980–2988 (2017)

28. Long, C., Hua, G.: Multi-class multi-annotator active learning with robust gaussian process for visual recognition. In: Proceedings of the IEEE International Conference on Computer Vision, pp. 2839–2847 (2015)

29. Long, C., Hua, G., Kapoor, A.: Active visual recognition with expertise estimation in crowdsourcing. In: Proceedings of the IEEE International Conference on Computer Vision, pp. 3000–3007 (2013)

30. Lovchinsky, I., et al.: Discrepancy ratio: evaluating model performance when even experts disagree on the truth. In: International Conference on Learning Representations (2019)

31. Ma, L., Stückler, J., Kerl, C., Cremers, D.: Multi-view deep learning for consistent semantic mapping with RGB-D cameras. In: 2017 IEEE/RSJ International Conference on Intelligent Robots and Systems, pp. 598–605. IEEE (2017)

32. Meng, H., Kleinsmith, A., Bianchi-Berthouze, N.: Multi-score learning for affect recognition: the case of body postures. In: D'Mello, S., Graesser, A., Schuller, B., Martin, J.-C. (eds.) ACII 2011. LNCS, vol. 6974, pp. 225–234. Springer, Heidelberg (2011). https://doi.org/10.1007/978-3-642-24600-5_26

33. Postels, J., Ferroni, F., Coskun, H., Navab, N., Tombari, F.: Sampling-free epistemic uncertainty estimation using approximated variance propagation. In: Proceedings of the IEEE/CVF International Conference on Computer Vision, pp. 2931–2940 (2019)

34. Rajpurkar, P., et al.: MURA: large dataset for abnormality detection in musculoskeletal radiographs. arXiv preprint arXiv:1712.06957 (2017)

35. Shen, Y., Zhang, Z., Sabuncu, M.R., Sun, L.: Real-time uncertainty estimation in computer vision via uncertainty-aware distribution distillation. In: Proceedings of the IEEE/CVF Winter Conference on Applications of Computer Vision, pp. 707–716 (2021)

36. Singh, A., et al.: Go-with-the-flow: tracking, analysis and sonification of movement and breathing to build confidence in activity despite chronic pain. Hum.-Comput. Interact. **31**(3–4), 335–383 (2016)

37. Surowiecki, J.: The Wisdom of Crowds. Anchor (2005)

38. Tanno, R., Saeedi, A., Sankaranarayanan, S., Alexander, D.C., Silberman, N.: Learning from noisy labels by regularized estimation of annotator confusion. In: Proceedings of the IEEE/CVF Conference on Computer Vision and Pattern Recognition, pp. 11244–11253 (2019)

39. Wang, C., Gao, Y., Mathur, A., Williams, A.C.D.C., Lane, N.D., Bianchi-Berthouze, N.: Leveraging activity recognition to enable protective behavior detection in continuous data. Proc. ACM Interact. Mob. Wearable Ubiquit. Technol. **5**(2) (2021)

40. Wang, C., Olugbade, T.A., Mathur, A., Williams, A.C.D.C., Lane, N.D., Bianchi-Berthouze, N.: Chronic pain protective behavior detection with deep learning. ACM Trans. Comput. Healthc. **2**(3), 1–24 (2021)

41. Warfield, S.K., Zou, K.H., Wells, W.M.: Simultaneous truth and performance level estimation (staple): an algorithm for the validation of image segmentation. IEEE Trans. Med. Imaging **23**(7), 903–921 (2004)

42. Yan, Y., et al.: Modeling annotator expertise: Learning when everybody knows a bit of something. In: Proceedings of the Thirteenth International Conference on Artificial Intelligence and Statistics, pp. 932–939. JMLR Workshop and Conference Proceedings (2010)
43. Yan, Y., Rosales, R., Fung, G., Subramanian, R., Dy, J.: Learning from multiple annotators with varying expertise. Mach. Learn. **95**(3), 291–327 (2014)
44. Zhang, L., et al.: Disentangling human error from the ground truth in segmentation of medical images. arXiv preprint arXiv:2007.15963 (2020)

Conformal Prediction Masks: Visualizing Uncertainty in Medical Imaging

Gilad Kutiel, Regev Cohen$^{(\boxtimes)}$, Michael Elad, Daniel Freedman, and Ehud Rivlin

Verily Life Sciences, South San Francisco, USA
`regevcohen@google.com`

Abstract. Estimating uncertainty in image-to-image recovery networks is an important task, particularly as such networks are being increasingly deployed in the biological and medical imaging realms. A recent conformal prediction technique derives per-pixel uncertainty intervals, guaranteed to contain the true value with a user-specified probability. Yet, these intervals are hard to comprehend and fail to express uncertainty at a conceptual level. In this paper, we introduce a new approach for uncertainty quantification and visualization, based on masking. The proposed technique produces interpretable image masks with rigorous statistical guarantees for image regression problems. Given an image recovery model, our approach computes a mask such that a desired divergence between the masked reconstructed image and the masked true image is guaranteed to be less than a specified risk level, with high probability. The mask thus identifies reliable regions of the predicted image while highlighting areas of high uncertainty. Our approach is agnostic to the underlying recovery model and the true unknown data distribution. We evaluate the proposed approach on image colorization, image completion, and super-resolution tasks, attaining high quality performance on each.

1 Introduction

Deep Learning has been successful in many applications, spanning computer vision, speech recognition, natural language processing, and beyond [12,13]. For many years, researchers were mainly content in developing new techniques that achieve unprecedented accuracy, without concerns for understanding the uncertainty implicit in such models. Recently, however, there has been a concerted effort within the research community to quantify the uncertainty of deep models.

This paper addresses the problem of quantifying and visualizing uncertainty in the realm of image-to-image tasks. Such problems include super-resolution, deblurring, colorization, and image completion, amongst others. Assessing uncertainty is important generally, but is particularly so in application domains such

---

Supplementary Information The online version contains supplementary material available at https://doi.org/10.1007/978-3-031-39539-0_14.

as biological and medical imaging, in which fidelity to the ground truth is paramount. If there is an area of the reconstructed image where such fidelity is unlikely or unreliable due to high uncertainty, this is crucial to convey.

Our approach to uncertainty estimation is based on masking. Specifically, we are interested in computing a mask such that the uncertain regions in the image are masked out. Based on conformal prediction [3], we derive an algorithm that can apply to any existing image-recovery model and produce uncertainty a mask satisfying the following criterion: the divergence between the masked reconstructed image and the masked true image is guaranteed to be less than a specified level, with high probability. The resultant mask highlights areas in the recovered image of high uncertainty while trustworthy regions remain intact. Our distribution-free method, illustrated in Fig. 1, is agnostic to the prediction model and to the choice of divergence function, which should be dictated by the application. Our contributions are as follows:

1. We introduce the notion of conformal prediction masks: a distribution-free approach to uncertainty quantification in image-to-image regression. We derive masks which visually convey regions of uncertainty while rigorously providing strong statistical guarantees for any regression model, image dataset and desired divergence measure.
2. We develop a practical training algorithm for computing these masks which only requires triplets of input (degraded), reconstructed and true images. The resultant mask model is trained once for all possible risk levels and is calibrated via a simple process to meet the required guarantees given a user-specified risk level and confidence probability.
3. We demonstrate the power of the method on image colorization, image completion and super-resolution tasks. By assessing our performance both visually and quantitatively, we show the resultant masks attain the probabilistic guarantee and provide interpretable uncertainty visualization without over-masking the recovered images, in contrast to competing techniques.

2   Related Work

Bayesian Uncertainty Quantification. The Bayesian paradigm defines uncertainty by assuming a distribution over the model parameters and/or activation functions. The most prevalent approach is Bayesian neural networks [21,28,38], which are stochastic models trained using Bayesian inference. Yet, as the number of model parameters has grown rapidly, computing the exact posteriors has became computationally intractable. This shortcoming has led to the development of approximation methods such as Monte Carlo dropout [15,16], stochastic gradient Markov chain Monte Carlo [11,33], Laplacian approximations [31] and variational inference [10,27,30]. Alternative Bayesian techniques include deep Gaussian processes [14], deep ensembles [8,19], and deep Bayesian active learning [17], to name just a few. A comprehensive review on Bayesian uncertainty quantification is given in [1].

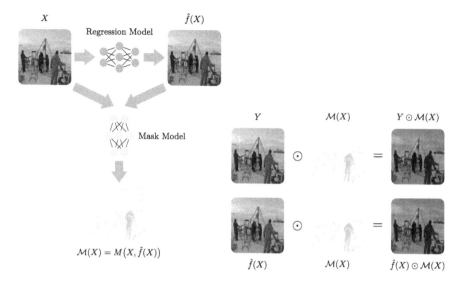

Fig. 1. High-level overview. Given image measurements X (e.g. gray-scale image) of a ground-truth image Y, and a predicted image $\hat{f}(X)$ (e.g. colorized image), the mask model outputs an uncertainty mask $\mathcal{M}(X)$ such that the divergence between the masked ground-truth and the masked prediction is below a chosen risk level with high probability.

Distribution-Free Methods and Conformal Prediction. Unlike Bayesian methods, the frequentist approach assumes the true model parameters are fixed with no underlying distribution. Examples of such distribution-free techniques are model ensembles [24,29], bootstrap [2,22], interval regression [23,29,39] and quantile regression [18,32]. An important distribution-free technique which is most relevant to our work is conformal prediction [4,36]. This approach relies on a labeled calibration dataset to convert point estimations into prediction regions. Conformal methods can be used with any estimator, require no retraining, are computationally efficient and provide coverage guarantees in finite samples [25]. Recent development includes conformalized quantile regression [7,32,35], conformal risk control [5,6,9] and semantic uncertainty intervals for generative adversarial networks [34]. [37] provides an extensive survey on distribution-free conformal prediction methods.

3   Background: Conformal Prediction in Image Regression

We present a brief overview of the work in [7], which stands out in the realm of conformal prediction for image-to-image problems, and serves as the basis of our work. Let $Y \in \mathcal{Y} = \mathbb{R}^N$ be a ground-truth image in vector form, an image $X \in \mathcal{X} = \mathbb{R}^M$ be its measurements, and $\hat{f}(X) \in \mathcal{Y}$ an estimator of Y. Conformal prediction constructs uncertainty intervals

$$\mathcal{T}(X)_{[i]} = \left[\hat{f}(X)_{[i]} - \hat{l}(X)_{[i]}, \hat{f}(X)_{[i]} + \hat{u}(X)_{[i]}\right], \quad i = 0, ..., N-1, \quad (1)$$

where $\hat{l}(X)_{[i]} \geq 0$ and $\hat{u}(X)_{[i]} \geq 0$ represent the uncertainty in lower and upper directions respectively. Given heuristic uncertainty values $\tilde{l}$ and $\tilde{u}$, the uncertainty intervals are calibrated using a calibration dataset $\mathcal{C} \triangleq \{X_k, Y_k\}_{k=1}^{K}$ to guarantee they contain at least a fraction α of the ground-truth pixel values with probability $1 - \delta$. Here $\alpha \in (0,1)$ and $\delta \in (0,1)$ are user-specified risk and error levels respectively. Formally, the per-pixel uncertainty intervals are defined as follows.

Definition 1. *Risk-Controlling Prediction Set (RCPS).* *A random set-valued function* $\mathcal{T} : \mathcal{X} \to \mathcal{Y}' = 2^{\mathcal{Y}}$ *is an* (α, δ)-*Risk-Controlling Prediction Set if*

$$\mathbb{P}(\mathcal{R}(\mathcal{T}) \leq 1 - \alpha) \geq 1 - \delta.$$

Here the risk is $\mathcal{R}(\mathcal{T}) \triangleq 1 - \mathbb{E}\left[\frac{1}{N}\left|\{i : Y_{[i]}^{test} \in \mathcal{T}(X^{test})_{[i]}\}\right|\right]$ *where the expectation is over a new test point* (X^{test}, Y^{test}), *while the outer probability is over the calibration data.*

The procedure for constructing RCPS consists of two stages. First, a machine learning system (e.g. neural network) is trained to output a point prediction $\hat{f}$, and heuristic lower and upper interval widths $(\tilde{l}, \tilde{u})$. The second phase utilizes the calibration set to calibrate $(\tilde{l}, \tilde{u})$ so they contain the right fraction of ground truth pixels. The final intervals are those in (1) with the calibrated widths $(\hat{l}, \hat{u})$.

Conformal prediction provides per-pixel uncertainty intervals with statistical guarantees in image-to-image regression problems. Yet, the per-pixel prediction sets may be difficult to comprehend on their own. To remedy this, the uncertainty intervals are visualized by passing the pixel-wise interval lengths through a colormap, where small sets render a pixel blue and large sets render it red. Thus, the redder a region is, the greater the uncertainty, and the bluer it is, the greater the confidence. The resultant uncertainty map, however, is not directly endowed with rigorous guarantees. This raises the following question: *can we directly produce an uncertainty map with strong statistical guarantees?*

4   Conformal Prediction Masks

Inspired by the above, we construct uncertainty masks $\mathcal{M}(X) = M(X, \hat{f}(X)) \in [0,1]^N$ such that

$$\mathbb{E}\left[\mathcal{M}(X^{test})_{[i]} \cdot \left|\hat{f}(X^{test})_{[i]} - Y_{[i]}^{test}\right|\right] \leq \beta_{[i]}, \tag{2}$$

where the expectation is over a new test point, and $\beta_{[i]} \in \mathbb{R}^+$ is user-specified risk level. Define $\hat{f}_{\mathcal{M}}(X) \triangleq \mathcal{M}(X) \odot \hat{f}(X)$ and $Y_{\mathcal{M}} \triangleq \mathcal{M}(X) \odot Y$ where $\odot$ represents a point-wise (Hadamard) product. Then, note that building (2) is equivalent to create the following uncertainty intervals

$$\mathcal{T}_{\mathcal{M}}(X)_{[i]} = [\hat{f}_{\mathcal{M}}(X)_{[i]} - \beta_{[i]}, \hat{f}_{\mathcal{M}}(X)_{[i]} + \beta_{[i]}], \tag{3}$$

which satisfies

$$Y^{\text{test}}_{\mathcal{M}[i]} \in \mathcal{T}_{\mathcal{M}}(X^{\text{test}})_{[i]}. \qquad (4)$$

We remark a few difference between (3) and (1): In (1) the lower and upper per-pixel uncertainty widths $(\hat{l}, \hat{u})$ depend on X and are calibrated, while in (3) $\hat{l} = \hat{u} \equiv \beta$ are user-specified and independent of X. Furthermore, the uncertainty parameters which undergo calibration are $\{\mathcal{M}(X)_{[i]}\}_{i=1}^{N}$.

One may notice that the above formulation exhibits a major limitation as each value of the prediction mask is defined independently from other values. Hence, it requires the user to specify a risk level for each pixel which is cumbersome, especially in high dimension. More importantly, setting each entry of the mask independently may fail in capturing the dependency between pixels, thus, fail to express uncertainty at a conceptual level. To overcome this, we redefine our uncertainty masks to ensure with probability at least $1 - \delta$ it holds

that $\mathbb{E}\left[\left\| \hat{f}_{\mathcal{M}}(X^{\text{test}}) - Y^{\text{test}}_{\mathcal{M}} \right\|_1 \right] \leq \alpha$, where $\alpha \in \mathbb{R}^+$ is a global risk level and

$\|Z\|_1 \triangleq \sum_{i=1}^{N} Z[i]$ is the L1 norm of an arbitrary image Z. Furthermore, the latter formulation can be generalized to any divergence measure $d : \mathcal{Y} \times \mathcal{Y} \to \mathbb{R}^+$ such that

$$\mathbb{E}\left[d\left(\hat{f}_{\mathcal{M}}(X^{\text{test}}), Y^{\text{test}}_{\mathcal{M}} \right) \right] \leq \alpha. \qquad (5)$$

Note we avoid trivial solutions, e.g. a zero-mask, which satisfy (5) yet provide no useful information. Thus, we seek solutions that employ the least masking required to meet (5), with high probability.

The above formulation enjoys several benefits. First, the current definition of the mask captures pixel-dependency. Thus, rather than focusing on individual pixels, the resultant map would mask out (or reduce) regions of high uncertainty within the predicated image to guarantee the divergence remains below the given risk level. Second, it accepts any divergence measure, each leading to a different mask. For example, selecting $d(\cdot, \cdot)$ to be a distortion measure may underline uncertainty regions of high-frequency objects (e.g. edges), while setting $d(\cdot, \cdot)$ to be a perceptual loss may highlight semantic factors within the image. Formally, we refers to these uncertainty masks as *Risk-Controlling Prediction Masks*, which are defined below.

Definition 2. *Risk-Controlling Prediction Mask (RCPM).* *A random function* $\mathcal{M} : \mathcal{X} \times \mathcal{Y} \to [0,1]^{\mathcal{Y}}$ *is an* (α, δ)-*Risk-Controlling Prediction Mask if*

$$\mathbb{P}\left(\mathbb{E}[\mathcal{R}(\mathcal{M})] \leq \alpha \right) \geq 1 - \delta,$$

where the risk is defined as $\mathcal{R}(\mathcal{M}) \triangleq d\left(\hat{f}_{\mathcal{M}}(X^{\text{test}}), Y^{\text{test}}_{\mathcal{M}} \right)$ *for given a divergence* $d(\cdot, \cdot)$. *The outer probability is over the calibration data, while the expectation taken over a test point* $(X^{\text{test}}, Y^{\text{test}})$.

As for RCPS, the procedure for creating RCPM includes two main stages. First, given a predictor $\hat{f}$, we require a heuristic notion of a non-zero uncertainty mask $\widetilde{M}$. In particular, we train a neural network to output a mask given the measurements and the predicted image as inputs. Second, given a divergence measure, we use the calibration set to calibrate the heuristic mask until the divergence measure decreases below the desired risk level. The final outputs are the calibrated mask and the original prediction multiplied by the mask. The overall method is outlined in Algorithm 1. Following the latter, we now discuss notion of initial uncertainty masks and the subsequent calibration process.

Algorithm 1. Generating RCPM

---

1. Given a regression model $\hat{f}$, train a model which outputs an initial mask $\widetilde{M}$.
2. Calibrate $\widetilde{M}$ using the calibration dataset to obtain $\mathcal{M}$ (e.g. using Algorithm 2).
3. Given X at inference, output the risk-controlling masked prediction $\hat{f}_{\mathcal{M}}(X) = \mathcal{M}(X) \odot \hat{f}(X)$.

4.1   Initial Estimation of Uncertainty Masks

Here we present two notions of uncertainty masks. The first concept, based on [7], translates given uncertainty intervals into a heuristic mask. In the second we develop a process for training a neural network which accepts the input and the predicted images and outputs an uncertainty mask based on a given divergence between the prediction and the ground-truth image.

Intervals to Masks. In [7], the authors propose to build uncertainty intervals based on four heuristic notions of lower and upper interval widths $\tilde{l}$ and $\tilde{u}$: (1) Regression to the magnitude of the residual; (2) one Gaussian per pixel; (3) softmax outputs; and (4) pixel-wise quantile regression. Then, we build a mask by setting the pixel-values to be inversely proportional to the interval sizes:

$$\widetilde{\mathcal{M}}(X)_{[i]} \propto \left(\tilde{u}_{[i]} - \tilde{l}_{[i]} \right)^{-1}. \tag{6}$$

Thus, the resultant mask holds high values at pixels with small-size intervals (high confidence) and smaller values at pixels with larger intervals corresponding to high uncertainty regions. However this approach requires first creating uncertainty intervals, hence, we next introduce a technique which directly produces an uncertainty mask.

Mask Regression. Here, we introduce a notion of an uncertainty mask represented by a neural network $\widetilde{\mathcal{M}}(X; \theta) \in [0,1]^N$ with parameters θ. The mask model is trained to output a mask which satisfies

$$\mathbb{E}\left[d\left(\hat{f}_{\widetilde{\mathcal{M}}}(X^{\text{train}}), Y^{\text{train}}_{\widetilde{\mathcal{M}}} \right) \right] \le \alpha. \tag{7}$$

where here the expectation is over the training samples $\mathcal{D} \triangleq \{X_j, Y_j\}_{j=1}^{J}$ used to train $\hat{f}$. To derive our loss function, we start with formulating the following problem for a given a triplet $(X, Y, \hat{f}(X))$

$$\min_{\theta} \|\widetilde{\mathcal{M}}(X, \hat{f}(X)) - \mathbb{1}\|_2^2 \quad \text{subject to} \quad d\left(\hat{f}_{\widetilde{\mathcal{M}}}(X), Y_{\widetilde{\mathcal{M}}}\right) \leq \alpha, \qquad (8)$$

where $\mathbb{1}$ is an image of all ones, representing no masking. The constraint in the above corresponds to (7), while the objective aims to find the minimal solution, i.e., the solution that masks the image the least (avoding trivial solutions). The Lagrangian of the problem is given by

$$\mathcal{L}(\theta, \mu) \triangleq \|\widetilde{\mathcal{M}}(X, \hat{f}(X)) - \mathbb{1}\|_2^2 + \mu\left(d\left(\hat{f}_{\widetilde{\mathcal{M}}}(X), Y_{\widetilde{\mathcal{M}}}\right) - \alpha\right) \qquad (9)$$

where $\mu > 0$ is the dual variable, considered as an hyperparameter. Given μ, the optimal mask can be obtained by minimizing $\mathcal{L}(\theta, \mu)$ with respect to θ, which is equivalent to minimizing

$$\|\widetilde{\mathcal{M}}(X, \hat{f}(X)) - \mathbb{1}\|_2^2 + \mu \cdot d\left(\hat{f}_{\widetilde{\mathcal{M}}}(X), Y_{\widetilde{\mathcal{M}}}\right) \qquad (10)$$

since α does not depend on θ. Thus, we train our mask model using the following loss function:

$$\mathcal{L}(\mathcal{D}, \theta) \triangleq \sum_{(X,Y)\in\mathcal{D}} \|\widetilde{\mathcal{M}}(X, \hat{f}(X)) - \mathbb{1}\|_2^2 + \mu \cdot d\left(\hat{f}_{\widetilde{\mathcal{M}}}(X), Y_{\widetilde{\mathcal{M}}}\right). \qquad (11)$$

The proposed approach facilitates the use of any differentiable distortion measure and is agnostic to the prediction model $\hat{f}$. Furthermore, notice that the loss function is independent of α, hence, can be trained once for all values of α. Thus, the output mask acts only as an initial uncertainty map which may not satisfy (5) and need to be calibrated. Following proper calibration, discussed next, our mask model attains (5) without requiring the ground-truth Y. Lastly, this approach directly outputs uncertainty masks and thus it is the focus of our work.

4.2   Mask Calibration

We consider the $\widetilde{M}(X)$ as an initial estimation of our uncertainty mask which needs to calibrated to provide the guarantee in Definition 2. As the calibration process is not the focus of our work, we perform a simple calibration outlined in Algorithm 2. The core of the calibration employs a parametric function $C(\cdot; \lambda)$ pixel-wise to obtain a mask $\mathcal{M}_\lambda(X)_{[i]} \triangleq C(\widetilde{\mathcal{M}}(X)_{[i]}; \lambda)$. In general, $C(\cdot; \lambda)$ can be any monotonic non-decreasing function. Here we consider the following form[1]

$$\mathcal{M}_\lambda(X)_{[i]} \triangleq \min\left(\frac{\lambda}{1 - \widetilde{\mathcal{M}}(X)_{[i]} + \epsilon}, 1\right) \quad \forall i = 1, ..., N, \qquad (12)$$

---

[1] A small value ϵ is added to the denominator to ensure numerical stability.

which has been found empirically to perform well in our experiments. To set $\lambda > 0$, we use the calibration dataset $\mathcal{C} \triangleq \{X_k, Y_k\}_{k=1}^K$ such that for any pair $(X_k, Y_k) \in \mathcal{C}$ we compute

$$\lambda_k \triangleq \max\left\{\hat{\lambda} : d\left(\hat{f}_{\mathcal{M}_{\hat{\lambda}}}(X_k), Y_{k\mathcal{M}_{\hat{\lambda}}}\right) \leq \alpha\right\}. \tag{13}$$

Finally, λ is taken to be the $1 - \delta$ quantile of $\{\lambda_k\}_{k=1}^K$, i.e. the maximal value for which at least δ fraction of the calibration set satisfies condition (5). Thus, assuming the calibration and test sets are i.i.d samples from the same distribution, the calibrated mask is guaranteed to satisfy Definition 2.

Algorithm 2. Calibration Process

Input: Calibration data $\mathcal{C} \triangleq \{X_k, Y_k\}_{k=1}^K$; risk level α; error rate δ; underlying predictor $\hat{f}$; heuristic mask $\widetilde{\mathcal{M}}$; a monotonic non-decreasing function $C(\cdot; \lambda) : [0,1] \to [0,1]$ parameterized by $\lambda > 0$.

1. For a given $\tilde{\lambda} > 0$, define $\mathcal{M}_{\tilde{\lambda}}(X)_{[i]} \triangleq C\left(\widetilde{\mathcal{M}}(X)_{[i]}; \tilde{\lambda}\right)$ for all $i = 1, ..., N$.
2. For each pair $(X_k, Y_k) \in \mathcal{C}$, set $\lambda_k \triangleq \max\left\{\hat{\lambda} : d\left(\hat{f}_{\mathcal{M}_{\hat{\lambda}}}(X_k), Y_{k\mathcal{M}_{\hat{\lambda}}}\right) \leq \alpha\right\}$.
3. Set λ to be the $1 - \delta$ quantile of $\{\lambda_k\}_{k=1}^K$.
4. Define the final mask model as $\mathcal{M}_\lambda(X)_{[i]} \triangleq C\left(\widetilde{\mathcal{M}}(X)_{[i]}; \lambda\right)$.

Output: Calibrated uncertainty mask model $\mathcal{M}_\lambda$.

---

5   Experiments

5.1   Datasets and Tasks

Datasets. Two data-sets are used in our experiments:

Places365 [40]: A large collection of 256×256 images from 365 scene categories. We use 1,803,460 images for training and 36,500 images for validation/test.

Rat Astrocyte Cells [26]: A dataset of 1,200 uncompressed images of scanned rat cells of resolution 990×708. We crop the images into 256×256 tiles, and randomly split them into train and validation/test sets of sizes 373,744 and 11,621 respectively. The tiles are partially overlapped as we use stride of 32 pixels when cropping the images.

Tasks. We consider the following image-to-image tasks (illustrated in Fig. 4 in the Appendix):

Image Completion: Using gray-scale version of Places365, we remove middle vertical and horizontal stripes of 32 pixel width, and aim to reconstruct the missing part.

Super Resolution: We experiment with this task on the two data-sets. The images are scaled down to 64×64 images where the goal is to reconstruct the original images.

Colorization: We convert the Places365 images to grayscale and aim to recover their colors.

5.2   Experimental Settings

Image-to-Image Models. We start with training models for the above three tasks. Note that these models are not intended to be state-of-the-art, but rather used to demonstrate the uncertainty estimation technique proposed in this work. We use the same model architecture for all tasks: an 8 layer U-Net. For each task we train two versions of the network: (i) A simple regressor; and (ii) A conditional GAN, where the generator plays the role of the reconstruction model. For the GAN, the discriminator is implemented as a 4 layer CNN. We use the L1 loss as the objective for the regressor, and add an adversarial loss for the conditional GAN, as in [20]. All models are trained for 10 epochs using Adam optimizer with a learning rate of 1e−5 and a batch size of 50.

Mask Model. For our mask model we use an 8 layer U-Net architecture for simplicity and compatibility with previous works. The input to the mask model are the measurement image and the predicated image, concatenated on the channel axis. The output is a mask having the same shape as the predicted image with values within the range $[0, 1]$. The mask model is trained using the loss function (11) with $\mu = 2$, a learning rate of 1e-5 and a batch size of 25.

Experiments. We consider the L1, L2, SSIM and LPIPS as our divergence measures. We set aside $1,000$ samples from each validation set for calibration and use the remaining samples for evaluation. We demonstrate the flexibility of our approach by conducting experiments on a variety of 12 settings: (i) Image Completion: {Regressor, GAN} × {L1, LPIPS}; (ii) Super Resolution: {Regressor, GAN} × {L1, SSIM}; and (iii) Colorization: {Regressor, GAN} × {L1, L2}.

Risk and Error Levels. Recall that given a predicted image, our goal is to find a mask that, when applied to both the prediction and the (unknown) reference image, reduces the distortion between them to a predefined risk level α with high probability δ. Here we fix $\delta = 0.9$ and set α to be the 0.1-quantile of each measure computed on a random sample from the validation set, i.e. roughly 10% of the predictions are already considered sufficiently good and do not require masking at all.

5.3   Competing Techniques for Comparison

Quantile – Interval-Based Technique. We compare our method to the quantile regression option presented in [7], denoted by Quantile. While their calibrated uncertainty intervals are markedly different from the expected distortion

172   G. Kutiel et al.

we consider, we can use these intervals and transform them into a mask using
(6). For completeness, we also report the performance of the quantile regression
even when it is less suitable, i.e. when the underlying model is a GAN and when
the divergence function is different from L1. We note again that for the sake of
a fair comparison, our implementation of the mask model uses exactly the same
architecture as the quantile regressor.

Opt – Oracle. We also compare our method with an oracle, denoted Opt,
which given a ground-truth image computes an optimal mask by minimizing
(10). We perform gradient descent using Adam optimizer with a learning rate
of 0.01, iterating until the divergence term decreases below the risk level α.
This approach is performed to each test image individually, thus no calibration
needed.

Comparison Metrics. Given a mask $\mathcal{M}(X)$ we assess its performance using
the following metrics: (i) Average mask size $s\big(\mathcal{M}(X)\big) \triangleq \frac{1}{N}\|\mathcal{M}(X) - \mathbb{1}\|_1$; (ii)
Pearson correlation $Corr(\mathcal{M}, d)$ between the mask size and the full (unmasked)
divergence value; and (iii) Pearson correlation $Corr(\mathcal{M}, \mathcal{M}_{opt})$ between the mask
size and the optimal mask $\mathcal{M}_{opt}$ obtained by Opt.

5.4   Results and Discussion

We now show a series of results that demonstrate our proposed uncertainty
masking approach, and its comparison with Opt and Quantile[2]. We begin with

Fig. 2. Examples of conformal prediction masks. The images from left to right
are the measurement, ground-truth, model prediction, our calibrated mask trained with
L1 loss and the ground-truth L1 error. Tasks are image completion (top), colorization
(middle) and super resolution (bottom).

---

[2] Due to space limitations, we show more extensive experimental results in the
Appendix, while presenting a selected portion of them here.

Table 1. Quantitative results. Arrows points to the better direction where best results are in **blue**.

		$s(\mathcal{M})$ ($\downarrow$)			$Corr(\mathcal{M},d)$ ($\uparrow$)		$Corr(\mathcal{M},\mathcal{M}_{opt})$ ($\uparrow$)	
Network	Distance	Opt	Ours	Quantile	Ours	Quantile	Ours	Quantile
Image Completion - Places365								
Regression	L1	0.09	0.10	0.15	0.89	0.78	0.89	0.76
Regression	LPIPS	0.01	0.01	0.20	0.54	0.51	0.89	0.77
GAN	L1	0.09	0.09	0.14	0.95	0.85	0.94	0.80
GAN	LPIPS	0.01	0.01	0.08	0.31	0.24	0.50	0.23
Super Resolution - Rat Astrocyte Cells								
Regression	L1	0.24	0.26	0.28	0.99	0.54	0.95	0.88
Regression	SSIM	0.03	0.03	0.13	0.66	0.64	0.82	0.57
GAN	L1	0.26	0.30	0.40	0.94	0.63	0.80	0.72
GAN	SSIM	0.03	0.03	0.13	0.79	0.63	0.83	0.63
Super Resolution - Places365								
Regression	L1	0.30	0.36	0.39	0.99	0.97	0.95	0.94
Regression	SSIM	0.10	0.23	0.48	0.89	0.85	0.94	0.84
GAN	L1	0.37	0.38	0.47	0.97	0.81	0.95	0.67
GAN	SSIM	0.10	0.12	0.51	0.86	0.81	0.92	0.86
Colorization - Places365								
Regression	L1	0.27	0.37	0.40	0.68	0.43	0.57	0.46
Regression	L2	0.18	0.37	0.38	0.57	0.30	0.60	0.48
GAN	L1	0.27	0.38	0.40	0.58	0.40	0.60	0.52
GAN	L2	0.18	0.36	0.38	0.42	0.28	0.59	0.49

a representative visual illustration of our proposed mask for several test cases in Fig. 2. As can be seen, the produced masks indeed identify sub-regions of high uncertainty. In the image completion task the bottom left corner is richer in details and thus there is high uncertainty regarding this part in the reconstructed image. In the colorization task, the mask highlights the colored area of the bus which is the most unreliable region since can be colorized with a large variety of colors. In the super resolution task the mask marks regions of edges and text while trustworthy parts such as smooth surfaces remain unmasked.

We present quantitative results in Table 1, showing that our method exhibits smaller mask sizes, aligned well with the masks obtained by Opt. In contrast, Quantile overestimates and produces larger masks as expected. In terms of the correlation $Corr(\mathcal{M},d)$, our method shows high agreement, while Quantile lags behind. This correlation indicates a much desired adaptivity of the estimated mask to the complexity of image content and thus to the corresponding uncertainty. We provide a complement illustration of the results in Fig. 3 in the Appendix. As seen from the top row, all three methods meet the probabilistic guarantees regarding the divergence/loss with fewer than 10% exceptions, as required. Naturally, Opt does not have outliers since each mask is optimally calibrated by its computation. The spread of loss values tends to be higher with Quantile, indicating weaker performance. The middle and bottom rows are consistent with results in Table 1, showing that our approach tends to produce masks that are close in size to those of Opt; while Quantile produces larger, and thus inferior, masked areas. We note that the colorization task seem to be more challenging, resulting in a marginal performance increase for our method compared to Quantile.

6   Conclusions

Uncertainty assessment in image-to-image regression problems is a challenging task, due to the implied complexity, the high dimensions involved, and the need to offer an effective and meaningful visualization of the estimated results. This work proposes a novel approach towards these challenges by constructing a conformal mask that visually-differentiate between trustworthy and uncertain regions in an estimated image. This mask provides a measure of uncertainty accompanied by an statistical guarantee, stating that with high probability, the divergence between the original and the recovered images over the non-masked regions is below a desired risk level. The presented paradigm is flexible, being agnostic to the choice of divergence measure, and the regression method employed.

References

1. Abdar, M., et al.: A review of uncertainty quantification in deep learning: techniques, applications and challenges. Inf. Fusion **76**, 243–297 (2021)
2. Alaa, A., Van Der Schaar, M.: Frequentist uncertainty in recurrent neural networks via blockwise influence functions. In: International Conference on Machine Learning, pp. 175–190. PMLR (2020)
3. Angelopoulos, A.N., Bates, S.: A gentle introduction to conformal prediction and distribution-free uncertainty quantification. CoRR abs/2107.07511 (2021). https://arxiv.org/abs/2107.07511
4. Angelopoulos, A.N., Bates, S.: A gentle introduction to conformal prediction and distribution-free uncertainty quantification. arXiv preprint arXiv:2107.07511 (2021)
5. Angelopoulos, A.N., Bates, S., Candès, E.J., Jordan, M.I., Lei, L.: Learn then test: calibrating predictive algorithms to achieve risk control. arXiv preprint arXiv:2110.01052 (2021)
6. Angelopoulos, A.N., Bates, S., Fisch, A., Lei, L., Schuster, T.: Conformal risk control. arXiv preprint arXiv:2208.02814 (2022)
7. Angelopoulos, A.N., et al.: Image-to-image regression with distribution-free uncertainty quantification and applications in imaging. arXiv preprint arXiv:2202.05265 (2022)
8. Ashukha, A., Lyzhov, A., Molchanov, D., Vetrov, D.: Pitfalls of in-domain uncertainty estimation and ensembling in deep learning. arXiv preprint arXiv:2002.06470 (2020)

9. Bates, S., Angelopoulos, A., Lei, L., Malik, J., Jordan, M.: Distribution-free, risk-controlling prediction sets. J. ACM (JACM) **68**(6), 1–34 (2021)
10. Blundell, C., Cornebise, J., Kavukcuoglu, K., Wierstra, D.: Weight uncertainty in neural network. In: International Conference on Machine Learning, pp. 1613–1622. PMLR (2015)
11. Chen, T., Fox, E., Guestrin, C.: Stochastic gradient Hamiltonian Monte Carlo. In: International Conference on Machine Learning, pp. 1683–1691. PMLR (2014)
12. Cohen, R., Blau, Y., Freedman, D., Rivlin, E.: It has potential: Gradient-driven denoisers for convergent solutions to inverse problems. Adv. Neural. Inf. Process. Syst. **34**, 18152–18164 (2021)
13. Cohen, R., Elad, M., Milanfar, P.: Regularization by denoising via fixed-point projection (red-pro). SIAM J. Imag. Sci. **14**(3), 1374–1406 (2021)
14. Damianou, A., Lawrence, N.D.: Deep gaussian processes. In: Artificial intelligence and statistics, pp. 207–215. PMLR (2013)
15. Gal, Y., Ghahramani, Z.: Dropout as a Bayesian approximation: representing model uncertainty in deep learning. In: International Conference on Machine Learning, pp. 1050–1059. PMLR (2016)
16. Gal, Y., Hron, J., Kendall, A.: Concrete dropout. Adv. Neural Inf. Process. Syst. **30** (2017)
17. Gal, Y., Islam, R., Ghahramani, Z.: Deep Bayesian active learning with image data. In: International Conference on Machine Learning, pp. 1183–1192. PMLR (2017)
18. Gasthaus, J., et al.: Probabilistic forecasting with spline quantile function RNNs. In: The 22nd International Conference on Artificial Intelligence and Statistics, pp. 1901–1910. PMLR (2019)
19. Hu, R., Huang, Q., Chang, S., Wang, H., He, J.: The MBPEP: a deep ensemble pruning algorithm providing high quality uncertainty prediction. Appl. Intell. **49**(8), 2942–2955 (2019). https://doi.org/10.1007/s10489-019-01421-8
20. Isola, P., Zhu, J.Y., Zhou, T., Efros, A.A.: Image-to-image translation with conditional adversarial networks. In: Proceedings of the IEEE Conference on Computer Vision and Pattern Recognition, pp. 1125–1134 (2017)
21. Izmailov, P., Maddox, W.J., Kirichenko, P., Garipov, T., Vetrov, D., Wilson, A.G.: Subspace inference for Bayesian deep learning. In: Uncertainty in Artificial Intelligence, pp. 1169–1179. PMLR (2020)
22. Kim, B., Xu, C., Barber, R.: Predictive inference is free with the jackknife+-after-bootstrap. Adv. Neural. Inf. Process. Syst. **33**, 4138–4149 (2020)
23. Kivaranovic, D., Johnson, K.D., Leeb, H.: Adaptive, distribution-free prediction intervals for deep networks. In: International Conference on Artificial Intelligence and Statistics, pp. 4346–4356. PMLR (2020)
24. Lakshminarayanan, B., Pritzel, A., Blundell, C.: Simple and scalable predictive uncertainty estimation using deep ensembles. Adv. Neural Inf. Process. Syst. **30** (2017)
25. Lei, J., G'Sell, M., Rinaldo, A., Tibshirani, R.J., Wasserman, L.: Distribution-free predictive inference for regression. J. Am. Stat. Assoc. **113**(523), 1094–1111 (2018)
26. Ljosa, V., Sokolnicki, K.L., Carpenter, A.E.: Annotated high-throughput microscopy image sets for validation. Nat. Methods **9**(7), 637–637 (2012)
27. Louizos, C., Welling, M.: Multiplicative normalizing flows for variational Bayesian neural networks. In: International Conference on Machine Learning, pp. 2218–2227. PMLR (2017)
28. MacKay, D.J.: Bayesian interpolation. Neural Comput. **4**(3), 415–447 (1992)

29. Pearce, T., Brintrup, A., Zaki, M., Neely, A.: High-quality prediction intervals for deep learning: a distribution-free, ensembled approach. In: International Conference on Machine Learning, pp. 4075–4084. PMLR (2018)

30. Posch, K., Steinbrener, J., Pilz, J.: Variational inference to measure model uncertainty in deep neural networks. arXiv preprint arXiv:1902.10189 (2019)

31. Ritter, H., Botev, A., Barber, D.: A scalable Laplace approximation for neural networks. In: 6th International Conference on Learning Representations, ICLR 2018-Conference Track Proceedings, vol. 6. International Conference on Representation Learning (2018)

32. Romano, Y., Patterson, E., Candes, E.: Conformalized quantile regression. Adv. Neural Inf. Process. Syst. **32** (2019)

33. Salimans, T., Kingma, D., Welling, M.: Markov chain Monte Carlo and variational inference: bridging the gap. In: International Conference on Machine Learning, pp. 1218–1226. PMLR (2015)

34. Sankaranarayanan, S., Angelopoulos, A.N., Bates, S., Romano, Y., Isola, P.: Semantic uncertainty intervals for disentangled latent spaces. arXiv preprint arXiv:2207.10074 (2022)

35. Sesia, M., Candès, E.J.: A comparison of some conformal quantile regression methods. Stat **9**(1), e261 (2020)

36. Shafer, G., Vovk, V.: A tutorial on conformal prediction. J. Mach. Learn. Res. **9**(3) (2008)

37. Sun, S.: Conformal methods for quantifying uncertainty in spatiotemporal data: a survey. arXiv preprint arXiv:2209.03580 (2022)

38. Valentin Jospin, L., Buntine, W., Boussaid, F., Laga, H., Bennamoun, M.: Hands-on Bayesian neural networks-a tutorial for deep learning users. arXiv e-prints pp. arXiv-2007 (2020)

39. Wu, D., et al.: Quantifying uncertainty in deep spatiotemporal forecasting. arXiv preprint arXiv:2105.11982 (2021)

40. Zhou, B., Lapedriza, A., Khosla, A., Oliva, A., Torralba, A.: Places: a 10 million image database for scene recognition. IEEE Trans. Pattern Anal. Mach. Intell. (2017)

Why Deep Surgical Models Fail?: Revisiting Surgical Action Triplet Recognition Through the Lens of Robustness

Yanqi Cheng[1(✉)], Lihao Liu[1], Shujun Wang[1], Yueming Jin[2],
Carola-Bibiane Schönlieb[1], and Angelica I. Aviles-Rivero[1]

[1] Department of Applied Mathematics and Theoretical Physics, University of Cambridge,
Cambridge, UK
{yc443,ll610,sw991,cbs31,ai323}@cam.ac.uk
[2] Wellcome/EPSRC Centre for Interventional and Surgical Sciences and Department
of Computer Science, University College London, London, UK
yueming.jin@ucl.ac.uk

Abstract. Surgical action triplet recognition provides a better understanding of the surgical scene. This task is of high relevance as it provides the surgeon with context-aware support and safety. The current go-to strategy for improving performance is the development of new network mechanisms. However, the performance of current state-of-the-art techniques is substantially lower than other surgical tasks. Why is this happening? This is the question that we address in this work. We present the first study to understand the failure of existing deep learning models through the lens of robustness and explainability. Firstly, we study current existing models under weak and strong $\delta-$perturbations via an adversarial optimisation scheme. We then analyse the failure modes via feature based explanations. Our study reveals that the key to improving performance and increasing reliability is in the core and spurious attributes. Our work opens the door to more trustworthy and reliable deep learning models in surgical data science.
https://yc443.github.io/robustIVT/.

Keywords: Robustness · Deep Learning · Surgical Triplet Recognition

1 Introduction

Minimally Invasive Surgery (MIS) has become the gold standard for several procedures (i.e., cholecystectomy & appendectomy), as it provides better clinical outcomes including reducing blood loss, minimising trauma to the body, causing less post-operative pain and faster recovery [28,30]. Despite the benefits of MIS, surgeons lose direct vision and touch on the target, which decreases surgeon-patient transparency imposing technical challenges to the surgeon. These challenges have motivated the development of automatic techniques for the analysis of the surgical workflow [2,14,18,29]. In particular, this work aims to address a key research problem in surgical data science– surgical recognition, which provides to the surgeon context-aware support and safety.

The majority of existing surgical recognition techniques focus on phase recognition [3,4,13,27,36]. However, phase recognition is limited by its own definition; as it does not provide complete information on the surgical scene. We therefore consider the setting of *surgical action triplet recognition*, which offers a better understanding of the surgical scene. The goal of triplet recognition is to recognise the ⟨instrument, verb, target⟩ and their inherent relations. A visualisation of this task is displayed in Fig. 1.

The concept behind triplet recognition has been recognised in the early works of that [9, 15]. However, it has not been until the recent introduction of richer datasets, such as CholecT40 [16], that the community started developing new techniques under more realistic conditions. The work of that Nwoye et al [16] proposed a framework called Tripnet, which was the first work to formally address surgical actions as triplets. In that work, authors proposed a 3D interaction space for learning the triplets. In more recent work, the authors of [18] introduced two new models. The first one is a direct extension of Tripnet called Attention Tripnet, where the novelty relies on a spatial attention mechanism. In the same work, the authors introduced another model called Rendezvous (RDV) that highlights a transformer-inspired neural network.

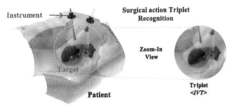

A commonality of existing surgical action triplet recognition techniques is the development of new mechanisms for improving the network architecture. However and despite the potential improvements, the performance of existing techniques is substantially lower than other tasks in surgical sciences– for example, force estimation and navigation assisted surgery. In this work, we go contrariwise existing techniques, and tackle the surgical action triplet recognition problem from the lens of robustness and explainability.

Fig. 1. Visualisation of the surgical action triplet recognition task. We consider the tasks where the instrument (*I*), verb (*V*, action), and target (*T*, anatomical part) seek to be predicted.

In the machine learning community there is a substantial increase of interest in understanding the lack of reliability of deep learning models (e.g., [7, 10, 11, 21, 26, 33]). To understand the lack of reliability of existing deep networks, a popular family of techniques is the so-called *feature based explanations* via robustness analysis [20, 23, 24, 31, 34]. Whilst existing techniques have extensively been evaluated for natural images tasks, there are no existing works addressing the complex problems as in action triplet recognition.

🖒 **Contributions.** In this work, we introduce, to the best of our knowledge, *the first study to understand the failure of existing deep learning models for surgical action triplet recognition*. To do this, we analyse the failures of existing state-of-the-art solutions through the lens of robustness. Specifically, we push to the limit the existing SOTA techniques for surgical action triplet recognition under weak and strong δ-perturbations. We then extensively analyse the failure modes via the evaluation criteria Robustness-*S*, which analyses the behaviour of the models through feature based explanations. Our study reveals the impact of core and spurious features for more robust models. Our study opens the door to more trustworthy and reliable deep learning models in surgical data science, which is imperative for MIS.

2 Methodology

We describe two key parts for Surgical action triplet recognition task: i) our experimental settings along with assumptions and ii) how we evaluate robustness via adversarial optimisation. The workflow of our work is displayed in Fig. 2.

2.1 Surgical Action Triplet Recognition

In the surgical action triplet recognition problem, the main task is to recognise the triplet IVT, which is the composition of three components during surgery: instrument (I), verb (V), and target (T) in a given RGB image $x \in \mathbb{R}^{H \times W \times 3}$.

Formally, we consider a given set of samples $\{(x_n, y_n)\}_{n=1}^{N}$ with provided labels $\mathcal{Y} = \{0, 1, .., C_{IVT} - 1\}$ for $C_{IVT} = 100$ classes. We seek then to predict a function $f : \mathcal{X} \mapsto \mathcal{Y}$ such that f gets a good estimate for the unseen data. That is, a given parameterised deep learning model takes the image x as input, and outputs a set of class-wise presence probabilities, in our case 100 classes, under the IVT composition, $Y_{IVT} \in \mathbb{R}^{100}$, which we call it the logits of IVT. Since there are three individual components under the triplet composition, within the training network, we also considered the individual component $d^* \in \{I, V, T\}$, each with class number C_{d^*} (i.e. $C_I = 6$, $C_V = 10$, $C_T = 15$). The logits of each component, $Y_{d^*} \in \mathbb{R}^{C_{d^*}}$, are computed and used within the network.

In current state-of-the-art (SOTA) deep models [16, 18], there is a communal structure divided into three parts: i) the feature extraction backbone; ii) the individual component encoder; and iii) the triplet aggregation decoder that associate the components and output the logits of the IVT triplet. More precisely, the individual component encoder firstly concentrates on the instrument component to output Class Activation Maps (CAMs $\in \mathbb{R}^{H \times W \times C_d}$) and the logits Y_I of the instrument classes; the CAMs are then associated with the verb and target components separately for their logits (Y_V and Y_T) to address the instrument-centric nature of the triplet.

The current SOTA techniques for surgical action triplet recognition focus on improving the components ii) & iii). However, the performance is still substantially lower than other surgical tasks. Our intuition behind such behaviour is due to the inherently complex and ambiguous conditions in MIS, which reflects the inability of the models to learn meaningful features. Our work is then based on the following modelling hypothesis.

> **Hypothesis 2.1: Deep Features are key for Robustness**
>
> *Deep surgical techniques for triplet recognition lacks reliability due to the ineffective features. Therefore, the key to boosting performance, improving trustworthiness and reliability, and understanding failure of deep models is in the deep features.*

Following previous hypothesis, we address the questions of– why deep triplet recognition models fail? We do that by analysing the feature based explanations via robustness. To do this, we consider the current three SOTA techniques for our study: Tripnet [16], Attention Tripnet, and Rendezvous [18]. Moreover, we extensively investigate the repercussion of deep features using four widely used backbones ResNet-18, ResNet-50 [6], DenseNet-121 [8], and Swin Transformer [12]. In the next section, we detail our strategy for analysing robustness.

2.2   Feature Based Explanations via Robustness

Our models of the triplet recognition output the logits of triplets composition, we then use it to select our predicted label for the classification result. We define the model from image x to the predicted label $\hat{y}$ as $f : \mathcal{X} \to \mathcal{Y}$, where $\mathcal{X} \subset \mathbb{R}^{H \times W \times 3}, \mathcal{Y} = \{0, 1, 2, ..., C_{IVT} - 1\}$.

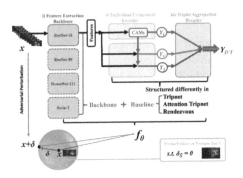

For each class $m \in \mathcal{Y}$ and within each given sample, we seek to recognise core and spurious attributions [24, 25], which definition is as follows.

Fig. 2. Illustration of the main network structure, and how the adversarial perturbation is added to measure robustness.

● **Core Attributes:** they refer to the features that form a part in the object we are detecting.

● **Spurious Attributes:** these are the ones that not a part of the object but co-occurs with it.

How We Evaluate Robustness? The body of literature has reported several alternatives for addressing the robustness of deep networks. Our work is motivated by recent findings on perturbation based methods, where even a small perturbation can significantly affect the performance of neural nets. In particular, we consider the setting of adversarial training [1,5,19] for *robustify* a given deep model.

The idea behind adversarial training for robustness is to enforce a given model to maintain its performance under a given perturbation δ. This problem can be seen cast as an optimisation problem over the network parameters θ as:

$$\theta^* = \arg\min_{\theta} \mathbb{E}_{(x,y) \sim \mathcal{D}} [\mathcal{L}_{\theta}(x, y)]. \tag{1}$$

where $\mathbb{E}[\mathcal{L}_{\theta}(\cdot)]$ denotes the expected loss to the parameter θ.

One seeks to the model be resistant to any δ−perturbation. In this work, we follow a generalised adversarial training model, which reads:

Definition 2.1: Adversarial training under δ

$$\theta^* = \arg\min_{\theta} \mathbb{E}_{(x,y) \sim \mathcal{D}} [\max_{\delta \in \Delta} \mathcal{L}_{\theta}(x + \delta, y)].$$

The goal is to the models do not change their performance even under the worse (strong) δ.

The machine learning literature has explored different forms of the generalised model in definition (2.1). For example, a better sparsity regulariser for the adversarial training as in [32]. In this work, we adopt the evaluation criteria of that [7], where one seeks to measure the susceptibility of features to adversarial perturbations. More precisely, we can have an insight of the deep features extracted by our prediction through visualising compact set of relevant features selected by some defined explanation methods on trained models, and measuring the robustness of the models by performing adversarial attacks on the relevant or the irrelevant features.

We denote the set of all features as U, and consider a general set of feature $S \subseteq U$. Since the feature we are interested are those in the image x, we further denote the subset of S that related to the image as x_S. To measure the robustness of the model, we rewrote the generalised model (2.1) following the evaluation criteria of that [7]. A model on input x with adversarial perturbation on feature set S then reads:

Definition 2.2: Adversarial δ & Robustness-S

$$\varepsilon_{x_S}^* := \{\min_{\delta} \|\delta\|_p \quad s.t. f(x+\delta) \neq y, \quad \delta_{\bar{S}} = 0\},$$

---

where y is the ground truth label of image x; $\|\cdot\|_p$ denotes the adversarial perturbation norm; $\bar{S} = U \setminus S$ denotes the complementary set of feature S with $\delta_{\bar{S}} = 0$ constraining the perturbation only happens on x_S. We refer to $\varepsilon_{x_S}^*$ as **Robustness-S** [7], or the minimum adversarial perturbation norm on x_S.

We then denote the relevant features selected by the explanation methods as $S_r \subseteq U$, with the irrelevant features as its complementary set $\bar{S_r} = U \setminus S_r$. Thus, the robustness on chosen feature sets– S_r and $\bar{S_r}$ tested on image x are:

$$\text{Robustness} - S_r = \varepsilon_{x_{S_r}}^*; \quad \text{Robustness} - \bar{S_r} = \varepsilon_{x_{\bar{S_r}}}^*.$$

3   Experimental Results

In this section, we describe in detail the range of experiments that we conducted to validate our methodology.

Table 1. Performance comparison for the task of Triplet recognition. The results are reported in terms of Average Precision (AP%) on the CholecT45 dataset using the official cross-validation split.

METHOD		COMPONENT DETECTION			TRIPLET ASSOCIATION		
BASELINE	BACKBONE	AP_I	AP_V	AP_T	AP_{IV}	AP_{IT}	AP_{IVT}
	ResNet-18	82.4 ± 2.5	54.1 ± 2.0	33.0 ± 2.3	30.6 ± 2.6	25.9 ± 1.5	21.2 ± 1.2
Tripnet	ResNet-50	85.3 ± 1.3	57.8 ± 1.6	34.7 ± 1.9	31.3 ± 2.3	27.1 ± 2.4	21.9 ± 1.5
	DenseNet-121	86.9 ± 1.4	58.7 ± 1.5	35.6 ± 2.8	33.4 ± 3.4	27.8 ± 1.8	22.5 ± 2.3
	ResNet-18	82.2 ± 2.6	56.7 ± 3.8	34.6 ± 2.2	30.8 ± 1.8	27.4 ± 1.3	21.7 ± 1.3
Attention Tripnet	ResNet-50	81.9 ± 3.0	56.8 ± 1.1	34.1 ± 1.4	31.5 ± 2.2	27.5 ± 1.0	21.9 ± 1.2
	DenseNet-121	83.7 ± 3.5	57.5 ± 3.2	34.3 ± 1.3	33.1 ± 2.4	28.5 ± 1.6	22.8 ± 1.3
	ResNet-18	85.3 ± 1.4	58.9 ± 2.6	35.2 ± 3.4	33.6 ± 2.6	30.1 ± 2.8	24.3 ± 2.3
Rendezvous	ResNet-50	85.4 ± 1.6	58.4 ± 1.4	34.7 ± 2.4	35.3 ± 3.5	30.8 ± 2.6	25.3 ± 2.7
	DenseNet-121	88.5 ± 2.7	61.7 ± 1.7	36.7 ± 2.1	36.5 ± 4.7	32.1 ± 2.7	26.3 ± 2.9
	Swin-T	73.6 ± 1.9	48.3 ± 2.6	29.2 ± 1.4	28.1 ± 3.1	24.7 ± 2.0	20.4 ± 2.1

Table 2. Heatmaps Comparison under different feature extraction backbones. We displayed four randomly selected images in fold 3 when using the best performed weights trained and validated on folds 1,2,4 and 5.

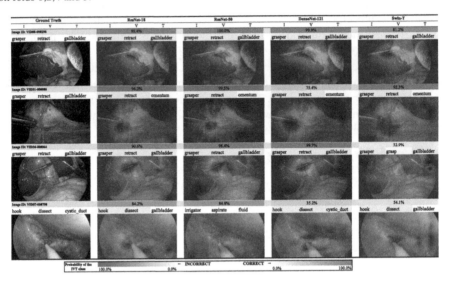

3.1 Dataset Description and Evaluation Protocol

Dataset Description. We use CholecT45 dataset [17] to evaluate the robustness of the three SOTA models for the Surgical Action Triplet Recognition task. Specifically, CholecT45 dataset contains 45 videos with annotations including 6 classes of instrument, 10 classes of verb, and 15 classes of target (i.e. $C_I = 6$, $C_V = 10$, $C_T = 15$) generating 900 ($6 \times 10 \times 25$) potential combinations for triplet labels. To maximise the clinical utility, we utilise the top-100 combinations of relevant labels, which are selected by removing a large portion of spurious combinations according to class grouping and surgical relevance rating [18]. Each video contains around 2,000 annotated frames extracted at 1 fps in RGB channels, leading to a total of 90,489 recorded frames. To remove the redundant information, the frames captured after the laparoscope been taken out of the body are blacked out with value $[0,0,0]$.

Evaluation Protocol. The triplet action recognition is evaluated by the average precision (AP) metric. Our models can directly output the predictions of triplet class AP_{IVT}. Instead, AP_d where $d \in \{I,V,T,IV,IT\}$ cannot be predicted explicitly. Then we obtain the final predictions of $d \in \{I,V,T,IV,IT\}$ components according to [17,18]:

$$Y_d^k = \max_m \{Y_{IVT}^m\}, \quad \forall m \in \{0,1..,C_{IVT}-1\} \, s.t. \, h_d(m) = k,$$

where we calculate the probability of class $k \in \{0,1,..,C_d-1\}$ under component d; and $h_d(\cdot)$ maps the class m from IVT triplet compositions to the class under component d.

In our robustness analysis, the main evaluation criteria is the robustness subject to the selected feature set (S_r and $\overline{S}_r$) on each backbone using the formula in (2.2).

Table 3. Top 5 predicted Triplet classes in each of the 10 models. The top 5 is assessed by the AP_{IVT} score.

	Tripnet		Attention Tripnet		Rendezvous	
	Triplet	AP	Triplet	AP	Triplet	AP
ResNet-18	12: grasper grasp specimen_bag	82.60%	12: grasper grasp specimen_bag	81.38%	17: grasper retract gallbladder	85.57%
	17: grasper retract gallbladder	81.04%	17: grasper retract gallbladder	78.70%	29: bipolar coagulate liver	83.90%
	29: bipolar coagulate liver	77.11%	29: bipolar coagulate liver	78.52%	12: grasper grasp specimen_bag	82.77%
	60: hook dissect gallbladder	74.13%	28: bipolar coagulate gallbladder	77.44%	30: bipolar coagulate omentum	76.88%
	79: clipper clip cystic_duct	61.28%	30: bipolar coagulate omentum	77.39%	60: hook dissect gallbladder	76.49%
ResNet-50	17: grasper retract gallbladder		17: grasper retract gallbladder	82.75%	30: bipolar coagulate omentum	91.36%
	12: grasper grasp specimen_bag	80.50%	12: grasper grasp specimen_bag	78.53%	17: grasper retract gallbladder	86.11%
	60: hook dissect gallbladder	77.15%	29: bipolar coagulate liver	76.44%	29: bipolar coagulate liver	84.94%
	29: bipolar coagulate liver	75.69%	60: hook dissect gallbladder	71.79%	12: grasper grasp specimen_bag	81.50%
	6: grasper grasp cystic_plate	69.24%	28: bipolar coagulate gallbladder	70.68%	28: bipolar coagulate gallbladder	79.60%
DenseNet-121	17: grasper retract gallbladder		17: grasper retract gallbladder	83.63%	84: irrigator dissect cystic_pedicle	96.84%
	12: grasper grasp specimen_bag	81.45%	12: grasper grasp specimen_bag	80.01%	30: bipolar coagulate omentum	89.60%
	29: bipolar coagulate liver	80.19%	29: bipolar coagulate liver	75.68%	17: grasper retract gallbladder	89.46%
	60: hook dissect gallbladder	76.35%	60: hook dissect gallbladder	75.36%	12: grasper grasp specimen_bag	85.88%
	79: clipper clip cystic_duct	67.75%	30: bipolar coagulate omentum	69.49%	29: bipolar coagulate liver	84.43%
Swin-T					17: grasper retract gallbladder	78.36%
					60: hook dissect gallbladder	72.57%
					12: grasper grasp specimen_bag	69.96%
					30: bipolar coagulate omentum	67.03%
					29: bipolar coagulate liver	66.08%

3.2 Implementation Details

We evaluate the model performance based on five-fold cross-validation, where we split 45 full videos into 5 equal folds. The testing set is selected from these 5 folds, and we treat the remaining 4 folds as the training set. Moreover, 5 videos from the 36 training set videos are selected as validation set during training.

The models are trained using the Stochastic Gradient Descent (SGD) optimiser. The feature extraction backbones are initialised with ImageNet pre-trained weights. Both linear and exponential decay of learning rate are used during training, with initial learning rates as $\{1e^{-2}, 1e^{-2}, 1e^{-2}\}$ for backbone, encoder and decoder parts respectively. We set the batch size as 32, and epoch which performs the best among all recorded epochs up to AP score saturation on validation set in the specified k-fold. To reduce computational load, the input images and corresponding segmentation masks are resized from 256×448 to 8×14. For fair comparison, we ran all SOTA models (following all suggested protocols from the official repository) under the same conditions and using the official cross-validation split of the CholecT45 dataset [17].

3.3 Evaluation on Downstream Tasks

In this section, we carefully analyse the current SOTA techniques for triplet recognition from the feature based explainability lens.

↻ **Results on Triplet Recognition with Cross-Validation.** As first part of our analysis, we investigate the performance limitation on current SOTA techniques, and emphasise how such limitation is linked to the lack of reliable features. The results are reported in Table 1. In a closer look at the results, we observe that ResNet-18, in general, performs the worst among the compared backbones. However, we can observe that for one case, component analysis, it performs better than ResNet-50 under Tripnet Attention baseline. The intuition being such behaviour is that the MIS setting relies on ambiguous condition and, in some cases, some frames might contain higher spurious features that are better captured by it. We remark that the mean and standard-deviation in Table 1 are calculated from the 5 folds in each combination of backbone and baseline.

We also observe that ResNet-50 performs better than ResNet-18 due to the deeper feature extraction. The best performance, for both the tasks– component detection and triplet association, is reported by DenseNet-121. The intuition behind the performance gain is that DenseNet-121 somehow mitigates the issue of the limitation of the capability representation. This is because ResNet type networks are limited by the identity shortcut that stabilises training. These results support our modelling hypothesis that the key of performance is the robustness of the deep features.

A key finding in our results is that whilst existing SOTA techniques [17, 18] are devoted to developing new network mechanisms, one can observe that a substantial performance improvement when improving the feature extraction. Moreover and unlike other surgical tasks, current techniques for triplet recognition are limited in performance. Why is this happening? Our results showed that the key is in the *reliable features* (linked to robustness); as enforcing more meaningful features, through several backbones, a significant performance improvement over all SOTA techniques is observed.

To further support our previous findings, we also ran a set of experiments using the trending principle of Transformers. More precisely, an non CNN backbone– the tiny Swin Transformer (Swin-T) [12] has also been tested on the Rendezvous, which has rather low *AP* scores on all of the 6 components in oppose to the 3 CNN backbones. This could be led by the shifted windows in the Swin-T, it is true that the shifted windows largely reduced the computational cost, but this could lead to bias feature attribute within bounding boxes, the incoherent spreading can be seen clearly in the visualisation of detected relevant features in Swin-T in Fig. 3(a).

In Table 1 we displayed the average results over all classes but– what behaviour can be observed from the per-class performance? It can be seen from Table 3 that though the best 5 predicted classes are different in each model, the predicted compositions seem clinically sensible supporting our previous discussion. In addition, the top 1 per-class *AP* score is significantly higher in DenseNet-121 with Rendezvous.

↻ **Visualisation Results.** To interpret features is far from being trivial. To address this issue, we provide a human-like comparison via heatmaps in Table 2. The implementation of the heatmaps is adapted from [35]. The displayed outputs reflect what the model is focusing based on the extracted features. These results support our hypothesis that deep features are the key in making correct predictions over any new network mechanism.

We observed that in the worst performed backbone– Swin-T, the feature been extracted are mostly spread across the images, however, the ones that concentrate on core attributes are not though performed the best. In the best performed DenseNet-121, a reasonable amount of attention are also been paid to spurious attributes; this can be seen more directly in our later discussion on robustness visualisation Fig. 3.

The reported probability on the predicted label emphasises again the outstanding performance of DenseNet-121 backbone; in the sense that, the higher the probability for the correct label the better, the lower it is for incorrect prediction the better.

⚡ **Why Surgical Triplet Recognition Models Fail? Robustness and Interpretability.** We further support our findings through the lens of robustness. We use as evaluation criteria Robustness-S_r and Robustness-$\overline{S_r}$ with different explanation methods: vanilla gradient (Grad) [22] and integrated gradient (IG) [26]. The results are in Table 4 & Fig. 3.

Table 4. Robustness measured on 400 examples (i.e. images) randomly selected from the images in the fold 3 videos with exactly 1 labeled triplet. Top 25 percent of relevant S_r or irrelevant $\overline{S_r}$ features are selected from 2 explanation methods Grad and IG. We perform attacks on the selected 25 percent.

ATTACKED FEATURES	EXPLANATION METHODS	BACKBONES (ON RENDEZVOUS)			
		ResNet-18	ResNet-50	DenseNet-121	Swin-T
Robustness-$\overline{S_r}$	Grad	2.599687	2.651435	**3.287798**	1.778592
	IG	2.621901	2.686064	**3.319311**	1.777737
Robustness-S_r	Grad	2.517404	2.608013	**3.188270**	1.750599
	IG	2.515343	2.603118	**3.187848**	1.749097

Comparison Between Different Backbones. In Table 4, we show the robustness results with top 25% attacked features on the average over 400 frames randomly chosen with exactly 1 labelled triplet. On one hand, we observe that the DenseNet-121 backbone consistently outperforms other network architectures on both evaluation criteria Robustness-S_r and Robustness-$\overline{S_r}$. This suggests that DenseNet-121 backbone does capture different explanation characteristics which ignored by other network backbones. On the other hand, our results are supported by the finding in [7], as IG performs better than Grad; and the attack on relevant features yields lower robustness than perturbing the same percentage of irrelevant features.

Robustness Explanation for Specific Images. To more objectively evaluate the robustness explanation for specific images, we show: (a) Visualisation of important features, (b) Robustness-$\overline{S_r}$, (c) Robustness against the percentage of Top features, and (d) Robustness-S_r in Fig. 3. In Fig. 3(a), we visualise the Top 15% features (with yellow dots) by Grad and IG, respectively, and overlay it on manually labelled region containing instrument (in red) and target (in green). We observe that the best performed backbone (can be seen from the robustness comparison curves in Fig. 3(c)) on the specific image is the one that not only pays attention to *core attributes, but also the spurious attribute*. In the image VID08-000188, the best performed model is ResNet-18, which shows the ambiguous condition on individual images. In a closer look at Fig. 3(a), a small portion of the most relevant feature extracted by ResNet-18 is spread not on the close surrounding of the object area. This importance of spurious attribute is further highlighted in image VID18-001156. We observe that DenseNet-121 provides the most

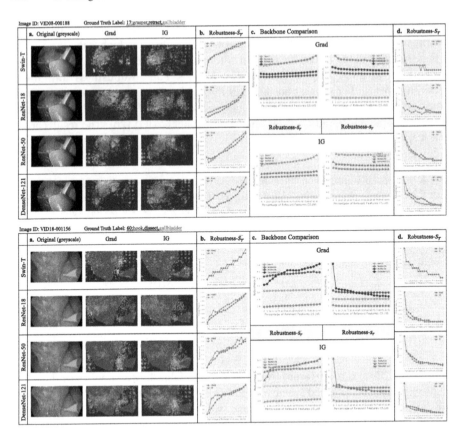

Fig. 3. The set of figures shows robustness analysis on randomly selected images with a. the visualisation of the Top 15 percent of important features selected by the 2 explanation methods-Grad and IG; b. (/d.) the trends showing the robustness measured on the relevant S_r (/irrelevant $\overline{S_r}$) features been selected by the 2 explanation methods against the percentage of Top features been defined as relevant; c. the comparison of the robustness across the 4 backbones embedded in Rendezvous baseline.

robust result highlighting relevant features within the tissue region and across tool tip. The worst performed model– ResNet-18 merely treated the core attributes as relevant.

The relevant role of spurious attributes can be explained by the nature of the triplet, which consists a verb component that is not the physical object. Overall, we observe that reliable deep features are the key for robust models in triplet recognition. Moreover, we observe, unlike existing works of robustness against spurious features, that both core and spurious attributes are key for the prediction.

4   Conclusion

We present the first work to understand the failure of existing deep learning models for the task of triplet recognition. We provided an extensive analysis through the lens of robustness. The significance of our work lies on understanding and addressing the key issues associated with the substantially limited in performance of existing techniques. Our work offers a step forward to more trustworthy and reliable models.

Acknowledgements. YC and AIAR greatly acknowledge support from a C2D3 Early Career Research Seed Fund and CMIH EP/T017961/1, University of Cambridge.

CBS acknowledges support from the Philip Leverhulme Prize, the Royal Society Wolfson Fellowship, the EPSRC advanced career fellowship EP/V029428/1, EPSRC grants EP/S026045/1 and EP/T003553/1, EP/N014588/1, EP/T017961/1, the Wellcome Innovator Awards 215733/Z/19/Z and 221633/Z/20/Z, the European Union Horizon 2020 research and innovation programme under the Marie Skodowska-Curie grant agreement No. 777826 NoMADS, the Cantab Capital Institute for the Mathematics of Information and the Alan Turing Institute.

References

1. Allen-Zhu, Z., Li, Y.: Feature purification: how adversarial training performs robust deep learning. In: 2021 IEEE 62nd Annual Symposium on Foundations of Computer Science (FOCS), pp. 977–988. IEEE (2022)
2. Aviles, A.I., Alsaleh, S.M., Hahn, J.K., Casals, A.: Towards retrieving force feedback in robotic-assisted surgery: a supervised neuro-recurrent-vision approach. IEEE Trans. Haptics **10**(3), 431–443 (2016)
3. Blum, T., Feußner, H., Navab, N.: Modeling and segmentation of surgical workflow from laparoscopic video. In: Jiang, T., Navab, N., Pluim, J.P.W., Viergever, M.A. (eds.) MICCAI 2010. LNCS, vol. 6363, pp. 400–407. Springer, Heidelberg (2010). https://doi.org/10.1007/978-3-642-15711-0_50
4. Dergachyova, O., Bouget, D., Huaulmé, A., Morandi, X., Jannin, P.: Automatic data-driven real-time segmentation and recognition of surgical workflow. Int. J. Comput. Assist. Radiol. Surg. **11**(6), 1081–1089 (2016). https://doi.org/10.1007/s11548-016-1371-x
5. Engstrom, L., Ilyas, A., Santurkar, S., Tsipras, D., Tran, B., Madry, A.: Adversarial robustness as a prior for learned representations. arXiv preprint arXiv:1906.00945 (2019)
6. He, K., Zhang, X., Ren, S., Sun, J.: Deep residual learning for image recognition (2015). https://doi.org/10.48550/ARXIV.1512.03385
7. Hsieh, C.Y., et al.: Evaluations and methods for explanation through robustness analysis. arXiv preprint arXiv:2006.00442 (2020)
8. Huang, G., Liu, Z., van der Maaten, L., Weinberger, K.Q.: Densely connected convolutional networks (2016). https://doi.org/10.48550/ARXIV.1608.06993
9. Katić, D., et al.: Knowledge-driven formalization of laparoscopic surgeries for rule-based intraoperative context-aware assistance. In: Stoyanov, D., Collins, D.L., Sakuma, I., Abolmaesumi, P., Jannin, P. (eds.) IPCAI 2014. LNCS, vol. 8498, pp. 158–167. Springer, Cham (2014). https://doi.org/10.1007/978-3-319-07521-1_17
10. Koh, P.W., Liang, P.: Understanding black-box predictions via influence functions. In: International Conference on Machine Learning, pp. 1885–1894. PMLR (2017)
11. Liu, L., Dou, Q., Chen, H., Qin, J., Heng, P.A.: Multi-task deep model with margin ranking loss for lung nodule analysis. IEEE Trans. Med. Imaging **39**(3), 718–728 (2019)
12. Liu, Z., et al.: Swin transformer: hierarchical vision transformer using shifted windows (2021). https://doi.org/10.48550/ARXIV.2103.14030

13. Lo, B.P.L., Darzi, A., Yang, G.-Z.: Episode classification for the analysis of tissue/instrument interaction with multiple visual cues. In: Ellis, R.E., Peters, T.M. (eds.) MICCAI 2003. LNCS, vol. 2878, pp. 230–237. Springer, Heidelberg (2003). https://doi.org/10.1007/978-3-540-39899-8_29

14. Maier-Hein, L., et al.: Surgical data science: enabling next-generation surgery. arXiv preprint arXiv:1701.06482 (2017)

15. Neumuth, T., Strauß, G., Meixensberger, J., Lemke, H.U., Burgert, O.: Acquisition of process descriptions from surgical interventions. In: Bressan, S., Küng, J., Wagner, R. (eds.) DEXA 2006. LNCS, vol. 4080, pp. 602–611. Springer, Heidelberg (2006). https://doi.org/10.1007/11827405_59

16. Nwoye, C.I., et al.: Recognition of instrument-tissue interactions in endoscopic videos via action triplets. In: Martel, A.L., et al. (eds.) MICCAI 2020. LNCS, vol. 12263, pp. 364–374. Springer, Cham (2020). https://doi.org/10.1007/978-3-030-59716-0_35

17. Nwoye, C.I., Padoy, N.: Data splits and metrics for method benchmarking on surgical action triplet datasets. arXiv preprint arXiv:2204.05235 (2022)

18. Nwoye, C.I., et al.: Rendezvous: attention mechanisms for the recognition of surgical action triplets in endoscopic videos. Med. Image Anal. 78, 102433 (2022)

19. Olah, C., et al.: The building blocks of interpretability. Distill 3(3), e10 (2018)

20. Plumb, G., Molitor, D., Talwalkar, A.S.: Model agnostic supervised local explanations. In: Advances in Neural Information Processing Systems, vol. 31 (2018)

21. Ribeiro, M.T., Singh, S., Guestrin, C.: "Why should I trust you?" explaining the predictions of any classifier. In: Proceedings of the 22nd ACM SIGKDD International Conference on Knowledge Discovery and Data Mining, pp. 1135–1144 (2016)

22. Shrikumar, A., Greenside, P., Kundaje, A.: Learning important features through propagating activation differences. In: International Conference on Machine Learning, pp. 3145–3153. PMLR (2017)

23. Simonyan, K., Vedaldi, A., Zisserman, A.: Deep inside convolutional networks: visualising image classification models and saliency maps. arXiv preprint arXiv:1312.6034 (2013)

24. Singla, S., Feizi, S.: Salient imagenet: how to discover spurious features in deep learning? In: International Conference on Learning Representations (2021)

25. Singla, S., Nushi, B., Shah, S., Kamar, E., Horvitz, E.: Understanding failures of deep networks via robust feature extraction. In: Proceedings of the IEEE/CVF Conference on Computer Vision and Pattern Recognition, pp. 12853–12862 (2021)

26. Sundararajan, M., Taly, A., Yan, Q.: Axiomatic attribution for deep networks. In: International Conference on Machine Learning, pp. 3319–3328. PMLR (2017)

27. Twinanda, A.P., Shehata, S., Mutter, D., Marescaux, J., De Mathelin, M., Padoy, N.: EndoNet: a deep architecture for recognition tasks on laparoscopic videos. IEEE Trans. Med. Imaging 36(1), 86–97 (2016)

28. Velanovich, V.: Laparoscopic vs open surgery. Surg. Endosc. 14(1), 16–21 (2000)

29. Vercauteren, T., Unberath, M., Padoy, N., Navab, N.: CAI4CAI: the rise of contextual artificial intelligence in computer-assisted interventions. Proc. IEEE 108(1), 198–214 (2019)

30. Wilson, E.B., Bagshahi, H., Woodruff, V.D.: Overview of general advantages, limitations, and strategies. In: Kim, K.C. (ed.) Robotics in General Surgery, pp. 17–22. Springer, New York (2014). https://doi.org/10.1007/978-1-4614-8739-5_3

31. Wong, E., Santurkar, S., Madry, A.: Leveraging sparse linear layers for debuggable deep networks. In: International Conference on Machine Learning, pp. 11205–11216. PMLR (2021)

32. Xu, K., et al.: Structured adversarial attack: towards general implementation and better interpretability. arXiv preprint arXiv:1808.01664 (2018)

33. Yeh, C.K., Hsieh, C.Y., Suggala, A., Inouye, D.I., Ravikumar, P.K.: On the (in) fidelity and sensitivity of explanations. In: Advances in Neural Information Processing Systems, vol. 32 (2019)

34. Zeiler, M.D., Fergus, R.: Visualizing and understanding convolutional networks. In: Fleet, D., Pajdla, T., Schiele, B., Tuytelaars, T. (eds.) ECCV 2014. LNCS, vol. 8689, pp. 818–833. Springer, Cham (2014). https://doi.org/10.1007/978-3-319-10590-1_53
35. Zhou, B., Khosla, A., Lapedriza, A., Oliva, A., Torralba, A.: Learning deep features for discriminative localization. In: Proceedings of the IEEE Conference on Computer Vision and Pattern Recognition, pp. 2921–2929 (2016)
36. Zisimopoulos, O., et al.: DeepPhase: surgical phase recognition in CATARACTS videos. In: Frangi, A.F., Schnabel, J.A., Davatzikos, C., Alberola-López, C., Fichtinger, G. (eds.) MICCAI 2018. LNCS, vol. 11073, pp. 265–272. Springer, Cham (2018). https://doi.org/10.1007/978-3-030-00937-3_31

Geometry-Based End-to-End Segmentation of Coronary Artery in Computed Tomography Angiography

Xiaoyu Yang[1,2], Lijian Xu[2,3(✉)], Simon Yu[4], Qing Xia[5], Hongsheng Li[6], and Shaoting Zhang[2]

[1] College of Electronics and Information Engineering, Tongji University, Shanghai, China
[2] Shanghai Artificial Intelligence Laboratory, Shanghai, China
[3] Centre for Perceptual and Interactive Intelligence, The Chinese University of Hong Kong, Hongkong, China
bmexlj@gmail.com
[4] Department of Imaging and Interventional Radiology, The Chinese University of Hong Kong, Hongkong, China
[5] SenseTime Research, Hongkong, China
[6] Department of Electronic Engineering, The Chinese University of Hong Kong, Hongkong, China

Abstract. Coronary artery segmentation has great significance in providing morphological information and treatment guidance in clinics. However, the complex structures with tiny and narrow branches of the coronary artery bring it a great challenge. Limited by the low resolution and poor contrast of medical images, voxel-based segmentation methods could potentially lead to fragmentation of segmented vessels and surface voids are commonly found in the reconstructed mesh. Therefore, we propose a geometry-based end-to-end segmentation method for the coronary artery in computed tomography angiography. A U-shaped network is applied to extract image features, which are projected to mesh space, driving the geometry-based network to deform the mesh. Integrating the ability of geometric deformation, the proposed network could output mesh results of the coronary artery directly. Besides, the centerline-based approach is utilized to produce the ground truth of the mesh instead of the traditional marching cube method. Extensive experiments on our collected dataset CCA-520 demonstrate the feasibility and robustness of our method. Quantitatively, our model achieves Dice of 0.779 and HD of 0.299, exceeding other methods in our dataset. Especially, our geometry-based model generates an accurate, intact and smooth mesh of the coronary artery, devoid of any fragmentations of segmented vessels.

Keywords: Segmentation · Coronary Artery · Geometry-based

E. Yang—This work is done during an internship at Shanghai Artificial Intelligence Laboratory.

H. Chen and L. Luo (Eds.): TML4H 2023, LNCS 13932, pp. 190–196, 2023.
https://doi.org/10.1007/978-3-031-39539-0_16

1   Introduction

Segmentation of the coronary artery tree in coronary computed tomography angiography (CCTA) is of great clinical value, such as presenting the morphology of the coronary artery, exhibiting the lesion and guiding clinical treatment. However, the automatic segmentation of the coronary artery indicates a severe challenge. First of all, the coronary artery has a distinctive tree structure with tiny and narrow branches that vary dramatically. Some branches are too thin to be segmented accurately, especially interfered with by other blood vessels. Second, the sparsity and anisotropy of CCTA images result in most segmentation methods being voxel-based. The reconstructed mesh from the voxel-based segmentation mask is rough with an obvious lattice shape. Additionally, limitations of CCTA images, such as low resolution and poor contrast, make it more challenging to segment the coronary artery.

Recently, deep learning has shown its viability of coronary artery segmentation with high accuracy. Most current methods perform voxel-based segmentation and achieve improvements based on the Unet, such as 3D-FFR-Unet [6], TETRIS [4], FFNet [12], PDS [10] and TreeConvGRU [3]. Instead of traditional voxel-based segmentation, mesh-deformation-based methods have been increasingly drawing the attention of the community. Voxel2Mesh [9] extends pixel2mesh [7] to 3D images for segmentation tasks of the liver, synaptic junction, and hippocampus. GMB [8] exploits point net to refine voxel-based segmentation results of the coronary artery by removing irrelevant vessels, where point cloud and voxel-based segmentation results are converted into each other.

Nonetheless, it remains a critical challenge to preserve the integrity and continuity of the coronary artery tree due to the existence of the fragmentation vessels. Additionally, the existing segmentation methods of integrating mesh deformation networks are limited to big organs with regular shapes, such as the liver and hippocampus. It is hard to generate complete and elaborate meshes of the coronary artery from voxel-based segmented results, due to its intricate structures and narrow branches.

To cope with the above problems, we propose a novel geometry-based segmentation network, where the generated vectorized mesh of the coronary artery becomes more integrated compared to the voxel-based segmentation results. Furthermore, the mesh results of the coronary artery are smoother with plentiful details, particularly in tiny and narrow branches. Extensive experiments confirm the robustness and feasibility of our method.

2   Methodology

Aiming at constructing the mesh of the coronary artery directly, an end-to-end cascade neural network is designed and illustrated as shown in Fig. 1. At stage I, a classical U-shaped neural network extracts image features in CCTA, which are projected into the mesh space. Guided by these projected features, the initial sphere mesh is deformed toward the ground truth through the graph

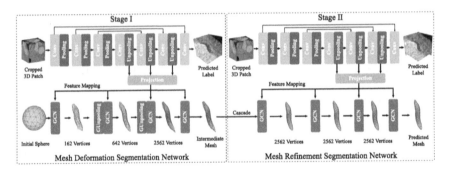

Fig. 1. Our geometry-based end-to-end cascade segmentation network for generating mesh of the coronary artery.

convolutional neural network (GCN). The mesh will add more vertices by graph unpooling to improve the details during the deformation. Besides, the segmentation neural network outputs voxel-based results and vectorized mesh results, simultaneously. The whole neural network is driven by voxel-based results and mesh results of the coronary artery. At stage II, refinement is performed on the vectorized mesh predicted by stage I. Without the need for graph unpooling, the GCN deforms the mesh to supplement more details of the coronary artery. The parameters of the U-shaped neural network are fixed and copied from the first stage, guaranteeing that the extracted image features are the same in both stages. The total loss consists of various losses to drive the U-shaped network and GCN as shown in Eq. 1.

$$\mathcal{L} = \lambda_1 \mathcal{L}_{Dice} + \lambda_2 \mathcal{L}_{CE} + \lambda_3 \mathcal{L}_{CD} + \lambda_4 \mathcal{L}_{Lap} + \lambda_5 \mathcal{L}_{NC} + \lambda_6 \mathcal{L}_{EG} \qquad (1)$$

where λ_{1-6} represents the weight of each loss. Dice and Cross-Entropy (CE) compute the loss of voxel-based results. Meanwhile, Chamfer Distance (CD) calculates the loss of point clouds between prediction and ground truth, and laplacian smoothing (Lap), normal consistency loss (NC) and edge loss (EG) are used to regularize it.

However, the intricate structure of the coronary artery exhibits a great challenge for the neural network. GCN is hard to learn such sophisticated and elaborate morphology. Accordingly, the cropped mesh of the coronary artery is classified into two categories: tube and bifurcation. Compared with twisted, irregular and multi-forks coronary artery mesh, tube and bifurcation have simpler morphology, which is more straightforward to be learned by the neural network. Hence, morphological regularization is devised to regularize cropped mesh into the structure of tube or bifurcation. Through regularized meshes, the geometry-based neural networks could learn the structural and morphological features of the coronary artery more easily and precisely.

On the other hand, we generate accurate ground truth meshes of the coronary artery by our centerline-based approach, instead of the traditional marching cube method. Along the centerline, obtained by skeletonizing the mask of the coronary artery, every branch of the mesh is reconstructed with a smooth and delicate surface. Given the mesh of each branch, the mesh boolean union operation is implemented to merge them and finish the complete mesh of the coronary artery. Our approach achieves the performance of reconstructing the ground truth meshes of the coronary artery, with integral structure and abundant details of tiny and narrow branches. In consequence, the reconstructed ground truth meshes of the coronary artery bring considerable improvement to the geometry-based segmentation neural network.

3   Experiments

In this section, the dataset and evaluation metrics are first introduced. Then, extensive experiments are conducted, evidencing the viability and practicality of coronary artery segmentation results generated by our model.

Table 1. Quantitative Evaluation Results of the Coronary Artery Segmentation for Different Methods on CCA-520 Dataset.

Methods	Dice	HD	Smoothness	Num of Segments	Chamfer Distance
ResUnet	0.575	3.960	0.550	116.4	111.85
H-DenseUnet	0.587	5.662	0.537	113.1	195.47
Unet3D	0.633	1.886	0.585	60.9	64.05
nnUNet	0.743	0.779	0.791	15.8	34.90
FFNet	0.707	6.026	0.729	126.6	29.80
3D-FFR-Unet	0.770	0.785	0.795	129.6	6.59
Voxel2Mesh	0.191	28.861	0.062	2.0	519.61
Ours	0.779	0.299	0.050	2.0	2.82

3.1   Dataset and Evaluation

Our proposed method is verified on our collected dataset CCA-520, which consists of 520 cases with coronary artery disease. To validate our model in small-scale data, comparative experiments are designed: 20 cases are used for training, and 500 cases for testing. Ground truth masks of 520 cases are coronary artery internal diameter annotations labelled by four radiologists.

Various metrics are applied to assess the performance of different models. Dice evaluates the intersection of segmentation results and ground truth. Hausdorff distance (HD) and chamfer distance measure the morphological difference. Smoothness judges the flatness of the segmented reconstruction mesh by computing the normal consistency for each pair of neighboring faces of the reconstruction mesh. Furthermore, for assessing the integrity and continuity of the coronary artery, the metric Num of Segments is proposed to count the number of connecting vessels.

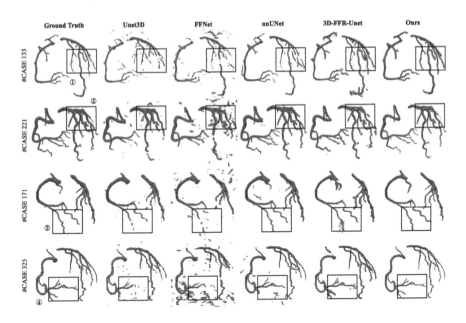

Fig. 2. Coronary Artery Segmentation Results on CCA-520 Dataset.① and ②: Complex multi-forks of the coronary artery. ③ and ④: Tiny and narrow branches

Several methods are selected for comparison experiments, namely ResUnet [11], H-DenseUnet [5], Unet3D [1], nnUNet [2], FFNet [12], 3D-FFR-Unet [6], Voxel2Mesh [9]. They belong to three main types of coronary artery segmentation, which are 2D pixel-based, 3D voxel-based and geometry-based segmentation methods, respectively.

3.2   Results and Discussion

Table 1 displays the quantitative evaluation results of the coronary artery segmentation for different methods on CCA-520 dataset. In terms of the similarity to ground truth, our method achieves Dice of 0.779 and hausdorff distance of 0.299, exceeding other segmentation methods of different types. The high Num of Segments of voxel/pixel-based segmentation methods indicates that they are severely interfered with by fragmentations of vessels, missing intact and complete structure of the coronary artery. The limitation of the sparsity also brings the disadvantage of low smoothness. Voxel2mesh predicts the organ as a whole, and the initial spheres are difficult to deform into complex branches of the coronary artery, resulting in a particularly low Dice. Faced with the complicated structure of the coronary artery, our approach can cope well with the segmentation task, generating accurate labels of the coronary artery. Besides, the geometry-based segmentation network outputs vectorized mesh of the coronary artery directly, bringing the highest smoothness in the reconstruction mesh of the coronary artery. Moreover, the geometry-based segmentation forms the mesh of the coronary artery by deforming initial spheres, so that no fragmentations of segmented vessels occur as with the voxel-based methods. The entire coronary artery tree is produced completely and elaborately. As for chamfer distance, it reveals our generated mesh of the coronary artery has similar morphology to the ground truth of the coronary artery with intricated morphology.

Figure 2 depicts the coronary artery segmentation results of different methods on our collected CCA-520 dataset. Voxel-based segmentation results mostly are hampered by fragmentations of vessels, missing an intact coronary artery structure. Conversely, our model produces meshes of the coronary artery with a complete structure, smooth multi-forks and clear tiny branches. The geometry-based segmentation network of the coronary artery elegantly avoids the fragmentations of segmented vessels and generates intact and continuous meshes of the coronary artery. Due to the vectorized characteristic of the mesh, the tiny and narrow branch end of the coronary artery can be more accurately delineated, eliminating the limitations of sparsity and the low resolution of CCTA images. By simplifying the training of our geometry-based segmentation network through morphological regularization, our model generates natural and smooth transitions at multi-forks of the coronary artery. In summary, the generated mesh of the coronary artery achieves superior performance in terms of accuracy, smoothness and integrity.

4 Conclusion

In this paper, we propose a geometry-based end-to-end segmentation model for the coronary artery tree with a complicated structure. The segmentation network is capable of generating precise, intact and smooth meshes, absent fragmentations of segmented vessels. Extensive experiments demonstrate our model, with a Dice of 0.779 on our CCA-520 dataset, surpassing other mainstream methods.

References

1. Çiçek, Ö., Abdulkadir, A., Lienkamp, S.S., Brox, T., Ronneberger, O.: 3D U-net: learning dense volumetric segmentation from sparse annotation. In: Ourselin, S., Joskowicz, L., Sabuncu, M.R., Unal, G., Wells, W. (eds.) MICCAI 2016. LNCS, vol. 9901, pp. 424–432. Springer, Cham (2016). https://doi.org/10.1007/978-3-319-46723-8_49
2. Isensee, F., Jaeger, P.F., Kohl, S.A.A., Petersen, J., Maier-Hein, K.H.: nnU-net: a self-configuring method for deep learning-based biomedical image segmentation. Nat. Meth. 18(2), 203–211 (2021)
3. Kong, B., et al.: Learning tree-structured representation for 3D coronary artery segmentation. Comput. Med. Imaging Graph. 80, 101688 (2020)
4. Lee, M.C.H., Petersen, K., Pawlowski, N., Glocker, B., Schaap, M.: TeTrIS: template transformer networks for image segmentation with shape priors. IEEE Trans. Med. Imaging 38(11), 2596–2606 (2019)

5. Li, X., Chen, H., Qi, X., Dou, Q., Fu, C.W., Heng, P.A.: H-DenseUNet: hybrid densely connected UNet for liver and tumor segmentation from CT volumes. IEEE Trans. Med. Imaging **37**(12), 2663–2674 (2018)

6. Song, A., et al.: Automatic coronary artery segmentation of CCTA images with an efficient feature-fusion-and-rectification 3D-UNet. IEEE J. Biomed. Health Inform. **26**(8), 4044–4055 (2022)

7. Wang, N., Zhang, Y., Li, Z., Fu, Y., Liu, W., Jiang, Y.G.: Pixel2Mesh: Generating 3D Mesh Models from Single RGB Images. In: Proceedings of the European Conference on Computer Vision (ECCV). pp. 52–67 (2018)

8. Wang, Q., et al.: Geometric morphology based irrelevant vessels removal for accurate coronary artery segmentation. In: 2021 IEEE 18th International Symposium on Biomedical Imaging (ISBI), pp. 757–760 (2021)

9. Wickramasinghe, U., Remelli, E., Knott, G., Fua, P.: Voxel2Mesh: 3D mesh model generation from volumetric data. In: Martel, A.L., et al. (eds.) MICCAI 2020. LNCS, vol. 12264, pp. 299–308. Springer, Cham (2020). https://doi.org/10.1007/978-3-030-59719-1_30

10. Zhang, X., et al.: Progressive deep segmentation of coronary artery via hierarchical topology learning. In: Wang, L., Dou, Q., Fletcher, P.T., Speidel, S., Li, S. (eds.) MICCAI 2022. LNCS, vol. 13435, pp. 391–400. Springer, Cham (2022). https://doi.org/10.1007/978-3-031-16443-9_38

11. Zhang, Z., Liu, Q., Wang, Y.: Road extraction by deep residual U-Net. IEEE Geosci. Remote Sens. Lett. **15**(5), 749–753 (2018)

12. Zhu, H., Song, S., Xu, L., Song, A., Yang, B.: Segmentation of coronary arteries images using spatio-temporal feature fusion network with combo loss. Cardiovasc. Eng. Technol. **13**(3), 407–418 (2022)

Author Index

H. Chen and L. Luo (Eds.): TML4H 2023, LNCS 13932, pp. 197–198, 2023.
https://doi.org/10.1007/978-3-031-39539-0

Printed in the United States
by Baker & Taylor Publisher Services